国外计算机科学教材系列

双语版C程序设计

(第2版)

Learn C through English and Chinese

Second Edition

[爱尔兰] Paul Kelly
苏小红 著

電子工業出版社
Publishing House of Electronics Industry
北京 • BEIJING

内 容 简 介

本书由在计算机程序设计方面有着丰富教学和实践经验的中外作者合作编写，并在其前一版的基础上进行了修订与更新。本书共14章内容，由浅入深全面介绍C程序设计方法，包括基本数据类型和基本输入/输出方式、各种控制结构和语句、指针和数组、字符串、函数、结构、文件输入和输出等内容，最后讨论了C预处理器。本书所有实例经过精心挑选、贴近生活，尤其强调读者的亲自参与意识。每章都为初学者提供了常见错误分析，所选习题可提高读者上机编程的兴趣。本书采用中英文对照混排，既方便初学者熟悉相关概念和内容，也便于英文非母语的读者熟悉英文专业词汇。

本书可作为高等学校计算机相关专业或软件学院的C程序设计双语教材，也可供程序员和编程爱好者参考使用。

图书在版编目（CIP）数据

双语版C程序设计：汉英对照/（爱尔兰）保罗•凯利（Paul Kelly），苏小红著. —2版.
北京：电子工业出版社，2017.3
国外计算机科学教材系列
ISBN 978-7-121-30555-9

Ⅰ.①双… Ⅱ.①保… ②苏… Ⅲ.①C语言－程序设计－高等学校－教材－汉、英
Ⅳ.①TP312.8

中国版本图书馆CIP数据核字（2016）第294634号

策划编辑：马 岚
责任编辑：冯小贝
印　　刷：涿州市京南印刷厂
装　　订：涿州市京南印刷厂
出版发行：电子工业出版社
　　　　　北京市海淀区万寿路173信箱　　邮编：100036
开　　本：787×1092　1/16　印张：13.75　字数：458千字
版　　次：2013 年 3 月第 1 版
　　　　　2017 年 3 月第 2 版
印　　次：2024 年 1 月第 9 次印刷
定　　价：45.00元

凡所购买电子工业出版社图书有缺损问题，请向购买书店调换。若书店售缺，请与本社发行部联系，联系及邮购电话：（010）88254888，88258888。
质量投诉请发邮件至zlts@phei.com.cn，盗版侵权举报请发邮件至dbqq@phei.com.cn。
本书咨询联系方式：classic-series-info@phei.com.cn。

Preface

Learn C through English and Chinese assumes no previous programming knowledge and has been carefully written to teach the fundamentals of programming through the medium of the C programming language. Although the book is primarily intended for use in a first year undergraduate Computer Science course, it will be equally useful to experienced programmers who intend to learn C programming on their own.

The book presents a thorough introduction to programming, using working programs to demonstrate the key features of the C language in a step-by-step logical sequence.

Using their professional teaching experience, the authors have carefully chosen each example program to carefully explain an important programming concept, confidently leading the learner to competence in C and programming in general. Emphasis is placed on key programming concepts and features of programming in C, while keeping explanations as simple and clear as possible.

Key concepts are re-enforced throughout the book by the use of 'Quick Syntax Reference' and 'Programming Pitfalls' sections at the end of each chapter. These are useful reference points for learners while writing their own programs.

Although mainly written in English, this book includes additional explanatory annotations in Chinese. This bi-lingual approach will be appreciated by Chinese learners and will help them focus on the learning task in hand, without being over-burdened by English technical terminology.

Despite C being available on a wide variety of platforms, this book is not specific to any particular machine, compiler, or operating system. All programs are designed to be portable with little or no modification to a wide variety of platforms.

Learn C through English and Chinese

- Is a comprehensive introduction to C programming.
- Uses practical examples to explain theoretical concepts.
- Uses an active learning approach with detailed explanations of working programming examples.
- Uses additional explanatory annotations written in Chinese.
- Each chapter contains
  - A 'Quick syntax reference' section.
  - A 'Programming pitfalls' section.
  - End-of-chapter exercises, allowing the learner to test and re-enforce their understanding of the programming concepts covered in the chapter.
- Is accompanied by a web site containing the example programs used in the book in addition to solutions to selected end-of-chapter exercises.

Typographic conventions used in this book.

The line numbers to the left of the program examples are for reference purposes only and are not part of the C language.

Keyboard input to a program is printed in this typeface.

When a C keyword is first introduce it is in bold and in this typeface.

`Program examples, screen output, reference to part of a C statement, a variable, a value in a program or a C keyword are in this typeface.`

When a new term is introduced it is in *italic* type.

前　　言

为了适应国内高等教育逐步与国际接轨的发展趋势，英语教学和双语教学越来越受到人们的重视。尤其是以“国际化、工业化”为办学理念并注重国际化、工业化人才培养的国家示范性软件学院的部分课程，还常常邀请一些外籍教师来国内进行全英语授课。但是由于办学经费有限，国内学生的英语水平又参差不齐，导致全英语教学目前还无法普及。因此，利用国内的优秀师资（包括留学回国人员）进行双语教学成为首选。针对目前许多双语课程教学缺少双语版教材的问题，我们组织国内和国外大学教师合作编写了本书。

本书的第一作者是爱尔兰都柏林工业大学（DIT）的高级讲师Paul Kelly。他长期从事程序设计类课程的教学工作，在程序设计类课程教学方面具有丰富的教学实践经验，在国外已先后出版多本程序设计语言类书籍，还曾数次被哈尔滨工业大学软件学院作为外聘教师应邀来校讲授程序设计方面的课程。Kelly对中国学生比较了解，针对其在教学中发现的问题，即初学者面临着既不熟悉专业术语和基本概念、又不熟悉英文专用词汇的双重困难，提出了出版中英文对照混排式双语版教材的思路，帮助学生在克服语言障碍的同时，能够更快、更好地熟悉和掌握计算机程序设计方面的基础知识，为国内的双语教学提供了一种最佳的解决方案。

本书内容共14章，由浅入深、全面介绍了C语言程序设计方法。本书的特点如下：

1. 使用实际生活中的例子和直观的图示，通俗易懂地讲解难于理解的概念。
2. 采用案例驱动和循序渐进的方式，从一个应用实例出发，先利用现有知识编写出一个较为简单的程序，然后在此基础上不断扩充，在扩充的过程中引入一个新的概念和知识点，逐渐编写出一个较大的程序。
3. 每个例程都有详细的讲解，重点内容和段落给出了中文注解，适合以C作为入门语言的读者对照阅读，既方便初学者熟悉相关概念和内容，也便于母语不是英语的读者熟悉英文的专业词汇，尤其适合双语教学。
4. 每章后面都有一节介绍初学者编程时易犯的错误，以帮助初学者在程序设计中避免这些错误。
5. 每章后面都有快速语法参考，总结本章内容，便于读者快速查询相关内容。
6. 每章后面都有精心设计的、有趣的习题，便于读者测试和强化对相关内容的理解。
7. 有相关的教学网站（华信教育资源网，网址http://www.hxedu.com.cn）和教材网站（http://book.sunner.cn），方便读者下载示例的源代码和教学课件等资料。

本书是继国内首次出版的中英文对照混排式双语版教材《双语版C++程序设计》之后出版的第二本双语版程序设计教材。Paul Kelly是一位治学非常严谨的教师，本书的第二作者苏小红在与他合著过程中，经常为一个细节内容的编写进行交流与讨论，书稿完成后又进行了多次校对工作。本着对所有读者负责的精神，我们真诚地欢迎读者对教材提出宝贵意见，可以通过发送电子邮件或在网站（http://book.sunner.cn）上留言等多种方式与我们交流讨论。此外，作者的电子邮件地址为Paul.Kelly@dit.ie 及sxh@hit.edu.cn。

苏小红

哈尔滨工业大学计算机科学与技术学院

目　　录

Chapter One
Introduction to C
第 1 章　引　　言

1.1　Brief history of C（C 语言简史）

The C programming language was originally developed at Bell Laboratories, New Jersey, USA in 1972. It was developed by Dennis Ritchie as a tool for writing the UNIX operating system. It evolved from a language called B, which in turn evolved from a language called BCPL (Basic Combined Programming Language).

C语言于1972年诞生在美国新泽西的贝尔实验室。最初，C语言是由Dennis Ritchie作为编写UNIX操作系统的工具而开发的，它是在B语言的基础上发展起来的，B语言是由一种称为BCPL（基本组合编程语言）的编程语言演变而来的。

The American National Standards Institute (ANSI) developed the first standardised specification for the language in 1989 and is commonly referred to as the C89 standard. The current ISO standard is C11, published in 2011.

1989 年，美国国家标准化协会（ANSI）开发了第一个C语言标准，通常称为C89。目前的ISO标准是2011年颁布的C11。

C is now widely used with many different operating systems running on many different hardware platforms. C, and other languages derived from it (such as Java, PHP and Perl), are widely used in Linux, Solaris, Macintosh, and Windows programming.

现在，C语言已经广泛运行于不同硬件平台的不同的操作系统中。C语言及由C语言派生的其他语言（如Java、PHP 和Perl），已广泛应用于基于Linux、Solaris、Macintosh 和Windows操作系统的程序设计中。

1.2　Why programmers use C（为什么程序员爱用C语言）

Here are just some of the reasons why programmers use C in preference to any other programming language:

相比其他编程语言，程序员更喜欢用C语言的几个原因如下。

1.2.1　C is portable

C is a portable language, which means that programs written in C on one computer system can, with little or no modifications, be transferred to and run on a completely different computer system. This should be a basic necessity for any programming language, but C comes nearest to meeting this important requirement.

C语言是可移植的编程语言，即在一种计算机系统中用C语言编写的程序在少量修改甚至无须修改的情况下就可以转换并运行于完全不同的其他计算机系统中。这是对所有编程语言的基本要求，但是，C语言是其中最能满足这个重要需求的一种语言。

C's portability is due to the adherence of C compilers to a specified standard. The specifications for C were originally laid down by Brian Kernighan and Dennis Ritchie in their book *The C Programming Language* (Prentice Hall, 1978). These specifications were further developed by ANSI, and most modern C compilers follow these specifications.

C语言的可移植性源于C编译器遵循了特定的标准。这些规范最初是由Brian Kernighan和Dennis Ritchie在其合著的*C Programming Language*中提出的。这些规范后来被美国国家标准化协会进一步完善，现在大部分现代的C编译器都遵循这个规范。

1.2.2 C is a structured programming language

C contains all the constructs required by modern structured programming techniques. The use of structured techniques results in programs that are reliable, understandable, and easy to maintain.

C语言包含现代结构化程序设计技术所需的所有结构，使用结构化程序设计技术可以编写可靠性高、可读性好、易于维护的程序。

1.2.3 C is efficient

Programs developed in C are smaller and faster than corresponding programs written in most other programming languages.

相比用其他大多数语言编写的程序，用C语言编写的程序更小、更快。

1.2.4 C is flexible

C can be used to write applications in a variety of areas, such as:

- artificial intelligence applications, such as expert systems and robotics
- commercial computer systems, such as accounts and distribution
- telecommunications applications
- computer-aided design (CAD)
- computer games
- database systems
- process control and real-time applications
- system programs, such as text editors, compilers, interpreters, and linkers.

  In fact C can be used to write a full operating system. For example, the major part of the Linux and Macintosh operating systems are written in C.
- mobile hand-held devices

C语言可用来编写各种应用领域中的程序，例如：

- 人工智能应用，例如专家系统和机器人
- 商业计算机系统，例如财务报表和销售
- 通信应用
- 计算机辅助设计（CAD）
- 计算机游戏
- 数据库系统
- 过程控制和实时应用
- 系统程序，例如文本编辑器、编译器、解释器和链接器

  事实上，C语言可以用来编写整个操作系统。例如，Linux和Macintosh操作系统的主要部分都是用C语言编写的。
- 移动手持设备

1.2.5 C is powerful

The C language contains some unique statements that allow for the efficient manipulation of system resources, such as memory, which would not be possible in other languages.

C语言包含一些独特的语句用于有效管理系统资源（例如内存），而其他编程语言是不具备这些功能的。

1.2.6　C is concise

Conciseness allows the programmer to give instructions to the computer in as few words as are necessary. Conciseness leads to smaller source programs, which are easier to edit and manage.

C语言的简洁性允许程序员只用所必需的只言片语就能向计算机发出指令，这使得用C编写的源程序更加短小，更易于编辑和管理。

1.3　Developing a C program（开发 C 程序）

Your first step in writing any computer program is to define and understand the problem to be solved. If you cannot understand the problem then you certainly will not be able to tell a computer how to solve the problem.

编写计算机程序的第一步是定义和理解所要解决的问题。如果不理解要解决的问题，那么你当然不能告诉计算机如何去解决这个问题。

The second step is to devise an *algorithm* to solve the problem. An algorithm is a series of well-defined steps that, if taken one by one, will solve the problem.

第二步是设计用于解决这个问题的算法，**算法**是一系列精心定义的解决问题的步骤，依次执行这些步骤就可以解决问题。

The third step is to design a program to implement the chosen algorithm.

第三步是设计程序，实现选定的算法。

These three steps are an extremely important stage of the development process and are similar in many ways to the work done by an architect before a house is built. Neglect at this stage will cause you to end up with something that does not suit your needs. Next comes the building stage, i.e. the actual writing of the program.

Firstly, the C program statements are typed and stored in a file using a *text editor*. The file containing the C statements is called the *source file* and is usually stored on disk.

上述三个步骤是程序开发过程中最为重要的阶段，与在建造房子前建筑师所做的工作有些类似。如果忽视了它们，那么将使程序最终在某些方面达不到用户的需求。接下来进入构建阶段，即实际编写程序。首先，使用**文本编辑器**将C语句键入并存储于某个文件中。这个包含C语句的文件称为**源文件**，通常存储在磁盘中。

Before you can run a C program on a computer you must pass the program through a *compiler*, which translates the C instructions to machine instructions that the computer ‘understands’. The compiler reads the source file, translates the C statements into machine or *object code*, and stores the object code in an *object file*.

在计算机上运行C程序之前，必须通过**编译器**将C 指令翻译成计算机能够“理解”的机器指令，编译器首先读取源文件，然后将C 语句翻译成机器代码或是**目标代码**，并将目标代码存储在**目标文件**中。

The final stage before running the program is to *link* the program using the *linker* program. Linking involves combining the object file of your program with other object files from the C run-time library to form an *executable file*.

运行程序前的最后一步是使用**链接器**对C程序进行链接，在链接过程中，将C程序的目标文件和来自于C运行库的其他目标文件进行组合，从而形成**可执行文件**。

The final step is to run the program and observe the results.

最后一步是运行程序，观察程序的输出结果。

In summary, the steps are:

1. Define the problem
↓
2. Design an algorithm to solve the problem
↓
3. Design the program
↓ output:
4. Write the code → source file
↓
5. Compile the source file → object file
↓
6. Link the object file → executable file
↓
7. Run the executable file → results

Figure 1.1 steps in developing a program

You are unlikely, except with the simplest of programs, to go from step 1 to step 7 without encountering some problem that will force you back to a previous step. For example, when you write the C statements you are likely to make some coding errors that the compiler will inform you of in step 5. These types of errors are called *syntax* errors. An example of a syntax error is the omission of a comma or semicolon in a C statement. To correct syntax errors go back to step 4, make the corrections, and return to step 5.

除非是编写最简单的程序，否则，在上述的步骤1~7中，常常会遇到一些问题，使你不得不返回前一个步骤。例如，在编辑C语句时，很可能出现代码输入错误，编译器在步骤5提示错误，这种类型的错误称为**语法**错误，例如，在C语言中少写了一个逗号或分号就是一个语法错误。为了修正语法错误就要回到步骤4，修正错误后，再继续执行步骤5。

Similarly, when you run the program you may find that it is not working as expected. Perhaps the C statements you wrote in step 4 do not implement the algorithm of step 2 correctly. This will involve you going back to step 4, but hopefully not any further. Going back to step 2 or 3 at this stage would be like asking an architect to redesign parts of the house while it is being built!

同样，在运行程序时，可能会发现程序并未按预期工作，这可能是因为在步骤4中编写的代码并不能正确地实现在步骤2中设计的算法。这将导致不得不回到步骤4修改代码，希望不是步骤2或步骤3的错误，因为如果是更前面步骤的问题，那么就类似于在建造房子时要求建筑师重新设计房子的部分结构一样。

Before being released for use, the program should be thoroughly tested to ensure it solves the problem defined in step 1.

在程序正式发布使用之前，必须对程序进行全面测试，以保证它能解决步骤1中定义的问题。

1.4 Suggestions for learning C programming （学习C 语言程序设计的建议）

This book encourages an active learning approach to gaining competence in programming. As you are reading, run the example programs and relate the program statements to the output from the programs. To save typing, the source code for all the example programs can be downloaded from the book's web site

本书鼓励读者采用积极主动的学习方法来提高自己的程序设计能力，一边阅读、一边上机运行程序并分析语句和程序输出结果之间的关系。

http://book.sunner.cn

or from

http://tinyurl.com/cprogs2e

Also, keep in mind that you'll learn more from designing, writing, running, and correcting programs than you ever will by simply reading the programs in the book. Practice is important.

To help you practise, there are exercises at the end of each chapter and there are links on the web site to free compilers and other resources. Do as many of the exercises as possible and get some feedback on your solutions from people who know C. However, do not rush into writing programs (step 4 in section 1.3) without giving steps 1 to 3 in section 1.3 careful consideration.

另外还要牢记，相对于单纯地阅读书上的程序而言，在设计、编写、运行和修改程序的过程中，你能学到更多。若要学好C语言，实践是很重要的。

为了帮助读者练习，每章都设置了一些练习题，并且可以通过前面提到的网站链接免费获取编译程序和其他资源。尽可能地多做练习题，然后向一些熟悉C语言的人请教问题的解决方法。但是建议读者不要急于编写程序，避免不经过前面讨论的步骤1~3的仔细分析就直接进入步骤4。

Chapter Two
C Data Types
第 2 章　C 数据类型

As in other programming languages, data can be of two types: *constants* and *variables*.

与其他编程语言一样，C语言中的数据分为两种类型：**常量和变量**。

2.1 Constants（常量）

As the name suggests, a constant does not change its value in a program. Some example of constants are shown in Table 2.1 below.

顾名思义，常量在程序运行过程中其值是保持不变的。

Table 2.1　some examples of constants

Type of constant	Examples	Remarks
Integer	`100, -3, 0`	Whole numbers that can be positive, negative or zero.
Float	`0.34, -12.34, 8.0`	Numbers with decimal parts.
Character	`'x', 'X', '*','1'`	Any character enclosed in single quotation marks.
String	`"abc", "A100", "1"`	One or more characters enclosed in double quotation marks.

2.2 Variables（变量）

Unlike a constant, a variable can vary its values in a program, and you must also define a variable before you can use it. A variable is defined by giving it a data type and a name.

与常量不同，变量在程序运行过程中其值是可以改变的。在使用变量前，必须先对变量进行定义。所谓定义变量也就是指定变量的类型和名称。

Program Example P2A

```
1  int main()
2  {
3    int v1 ;
4    float v2 ;
5    char v3 ;
6    v1 = 65 ;
7    v2 = -18.23 ;
8    v3 = 'a';
9    return 0 ;
10 }
```

This is an example of a C program.
The line numbers on the left are for reference only and are not part of the program.

这是一个C程序的实例，位于左侧的行号只是为了引用方便，它并不是程序的一部分。

The program starts with the line

```
int main()
```

This line marks the point where a C program starts to execute and must appear once only in a C program.

The program statements are contained within the braces { and } on lines 2 and 10. Each statement ends with a semicolon (;). The spaces before the semicolon and on each side of the equals sign are not essential and are used here only to improve the readability of the program.

Lines 3, 4 and 5 of this program define three variables: `v1`, `v2`, and `v3`. You can give a variable any name you wish, provided you keep within the following four rules:

1. A variable name can only be constructed using letters(a-z, A-Z), numerals (0-9) or underscores(_).
2. A variable name must start with a letter or an underscore.
3. A variable name cannot be a C *keyword*. A keyword is a word that has a special meaning. See appendix A for a list of keywords.
4. A variable name can contain any number of characters, but only the first thirty-one characters are significant to the C compiler.

The valid and invalid variable names are shown in Table 2.2.

Table 2.2 valid and invalid variable names

Name	Remarks
`month_1_sales`	This is a valid variable name.
`month1sales`	Valid, but not as readable as `month_1_sales`.
`1st_month_sales`	Invalid. The name does not start with a letter or an underscore.
`month 1 sales%`	Invalid. Spaces and % are not allowed.
`Xyz`	Valid, but variable names should be meaningful.
`SalesForThisWeek`	Valid and meaningful.
`int`	Invalid. This is a keyword.

Lines 3 to 5 of program P2A define `v1` as an integer variable, `v2` as a floating-point (can hold decimals) variable, and `v3` as a character variable. Note that variable names are case-sensitive, i.e. the variable `V1` is different from the variable `v1`.

Lines 6 to 8 of the program assign values to the variables. The value assigned to each variable is stored in the computer's memory.

The `return` statement on line 9 terminates the execution of the program and returns the integer `0` to the operating system. The `0` value is an indication to the operating system that the program executed successfully with no errors.

C程序是以int main( )行开始的，它标志着程序执行的入口点，在程序中只能出现一次。程序通过花括号{ }将第2~10行之间的语句括起来，每条语句均以分号结束，分号前面和等号两边的空格不是必需的，它们仅仅是为了提高程序的可读性。

程序第3、4、5行语句定义了三个变量v1、v2、v3。在遵守以下规则的前提下，可以根据自己的喜好对其随意命名。

1. 变量名只能由英文字母（a~z，A~Z）、数字（0~9）和下画线（_）构成。
2. 变量名只能以字母或下画线开头。
3. 不能用C语言的关键字对变量命名。**关键字**是系统定义的具有特殊含义的单词，参见附录A中的关键字列表。
4. 变量名可以包含任意数量的字符，但对于C编译器来说，只有前31个字符才是有意义的。

程序P2A的第3~5行定义了整型变量v1、浮点型（可以带小数）变量v2、字符型变量v3。注意，变量名是大小写敏感的，例如变量V1与变量v1是不同的变量。

程序的第6~8行为变量赋值，每个变量的值都存储在内存中。

第9行的return语句用于结束程序的执行，并返回一个整数0给操作系统。返回给操作系统的0值表示程序是正常结束的，没有出现错误。

2.3 Simple output to the screen （简单的屏幕输出）

Now that you have assigned values to the variables, how do you display their values on the screen? You can do this with **printf** (pronounced print-f) as shown in the next program.

既然已经对变量进行了赋值操作，那么如何将其显示到屏幕上呢？答案是使用printf。

Program Example P2B

```
1  #include <stdio.h>
2  int main()
3  {
4    int v1 ;
5    float v2 ;
6    char v3 ;
7    v1 = 65 ;
8    v2 = -18.23 ;
9    v3 = 'a' ;
10   printf( "v1 has the value %d\n", v1 ) ;
11   printf( "v2 has the value %f\n", v2 ) ;
12   printf( "v3 has the value %c\n", v3 ) ;
13   printf( "End of program\n" ) ;
14   return 0 ;
15 }
```

When you compile and run this program you will get the following on your screen:

```
v1 has the value 65
v2 has the value -18.230000
v3 has the value a
End of program
```

Line 1 is an example of a *preprocessor directive*. This line will be in nearly every program that you write. (Preprocessor directives are covered in Chapter Fourteen.)

程序第1行是一条**预处理指令**，这一行几乎出现在每个程序中（预处理指令将在第14章介绍）。

printf() is a standard library function for displaying data on the screen. The `printf()` functions in lines 10 to 12 have a string of characters enclosed in double quotation marks (called the *format string*) followed by a comma and a variable name. The variable name is not in quotation marks.

函数printf( )是一个用于在屏幕上显示数据的标准库函数，程序第10~12行的printf( )的括号中都有一个用双引号括起来的字符串，称为**格式字符串**，双引号后面是一个逗号，接着是变量名，变量名不能放到格式字符串所在的双引号中。

Let's look at the first `printf()` in line 10 and see how it works. All the characters between the double quotation marks up to the `%` symbol are displayed. The screen will then show:

双引号中位于%之前的字符都被显示到屏幕上。

```
v1 has the value
```

The letter `d` is called a conversion character and is preceded by a `%` sign. The `%d` (`%i` can also be used) displays the

字符d称为转换字符，以其前面的%作为其标识，%d（或者用%i）用于以十进制整型

variable v1 as a decimal integer on the screen. The variable v1 has the value 65, so the entire printf() on line 10 displays:

格式将v1显示到屏幕上。

```
v1 has the value 65
```

The newline character (\n) at the end of the format string is an instruction to skip to the next line of the screen.

格式化字符串最后的\n为换行符，其作用是将光标跳转到屏幕下一行的起始位置。

In the printf() functions on lines 11 and 12, the %f displays v2 as a floating-point number and the %c displays v3 as a character.

在第11行和第12行的printf( )中，使用%f以浮点型格式输出v2的值，使用%c以字符型格式输出v3 的值。

It is not always necessary to have a variable in a printf(). The printf() on line 13 has no variable and simply displays the line:

printf( )中的变量不是必须有的，如第13行的printf( )中就没有变量，只是简单地输出字符串。

```
End of program
```

2.4 Comments（注释）

Comments are added to a C program to make it more readable for the programmer, but they are completely ignored by the computer. Comments start with the characters /* and end with the characters */. We can add some comments to the last program.

在C程序中添加注释可以增加程序的可读性。但是在程序执行的过程中，计算机是完全忽略它们的。C语言中的注释以/*开始、以*/结束。

Program Example P2C

```
1  /* Program Example P2C
2     Introduction to variables in C. */
3  #include <stdio.h>
4  int main()
5  {
6    int v1 ;    /*  v1 is an integer variable.        */
7    float v2 ; /*  v2 is a floating-point variable. */
8    char v3 ;   /*  v3 is a character variable.       */
9    /* Now assign some values to the variables. */
10   v1 = 65 ;
11   v2 = -18.23 ;
12   v3 = 'a' ;
13   /* Finally display the variable values on the screen. */
14   printf( "v1 has the value %d\n", v1 ) ;
15   printf( "v2 has the value %f\n", v2 ) ;
16   printf( "v3 has the value %c\n", v3 ) ;
17   printf( "End of program\n" ) ;
18   return 0 ;
19 }
```

Comments can be placed anywhere in a C program and can span more than one line. Just make sure a comment starts with /* and ends with */.

注释可以放置在程序中的任何位置，并且可以跨越多行，只要确保注释的内容是以/*开始、以*/结束即可。

Comments cannot be *nested*, i.e. you cannot have a comment within another comment.

注释不可以嵌套，即不能在一个注释中添加另一个注释。

Typically, comments are placed at the start of the program to describe the purpose of the program, the author, date written and any other relevant information, such as the version number. For example:

通常在程序的开始位置添加注释，用于说明程序的用途、作者、编写的日期及其他一些相关的信息，如版本号等。

```
/*  Program name  : P2A.
    Introduction to variables in C.
    Written by    : Paul Kelly and Su Xiaohong.
    Date          : June 29, 2012.
    Version number: 1.0                      */
```

Comments are also used to describe in plain language the function of a particular section of a program. Get into the habit of using good explanatory comments; the more complicated the program becomes, the more valuable they will be to you and any other programmer reading your program.

注释还可用于通过简洁的文字来说明某段程序要实现的功能。要养成在程序中添加良好注释的习惯，程序越复杂，注释对阅读程序的程序员而言价值就越大。

2.5 Data types（数据类型）

In previous programs, it was shown how to declare a variable and associate it with a particular data type (`char`, `int` or `float`). The C language has a variety of other data types besides the three basic types of `char`, `int` and `float`. Different data types require different amounts of memory and therefore vary in the range of values they can store.

Details of the various data types in the C language are given in appendix D.

在前面的程序中，我们已经看到如何声明变量，如何将变量与一种特定的数据类型（如char、int、float）关联起来。除了char、int、float这三种基本的数据类型之外，C语言还提供了许多其他的数据类型。不同的数据类型占据不同大小的内存空间，因此也拥有不同的取值范围。

2.5.1 Short integer data types

A short integer has a smaller range of values than an integer and consequently uses less memory.

短整型能表示的数值范围小于基本整型，相应地短整型占用的内存也更少。

The following defines a variable `v1` as a `short integer`:

```
short int v1 ;
```

The keyword `int` is optional, so `v1` can also be defined as:

在这里，关键字int是可选的，所以v1也可以定义为如下形式。

```
short v1 ;
```

A `short int` variable, like an `int` variable, is displayed using `%d` as the format specifier in `printf()`. For example:

和基本整型变量一样，使用printf( )函数输出短整型变量的值，同样也使用%d格式转换符。

```
printf( "%d", v1 ) ;
```

2.5.2 Long integer data types

A long integer has a larger range of values than an integer. The following defines a variable v2 as a long integer:

长整型比基本整型的取值范围大。

```
long int v2 ;
```

The keyword int is optional here, so v2 can also be defined as:

```
long v2 ;
```

To display a long int variable, the format specifier %ld is used in printf(). For example:

使用printf( )函数输出长整型变量的值，应使用%ld格式转换符。

```
printf( "%ld", v2 ) ;
```

2.5.3 Unsigned integer data types

The keyword unsigned extends the positive range of an integer variable but does not allow negative values to be stored.
The following defines unsigned integer variables v3 and v4:

关键字unsigned扩展了整型变量可以表达的正整数的范围，但同时也使其不能再存储负数。

```
unsigned int v3 ;
unsigned long int v4 ;
```

To display these variables, %u and %lu are used as the format specifiers in printf(). For example:

为了利用printf( )函数输出无符号整型变量的值，应使用%u和%lu格式转换符。

```
printf( "%u %lu", v3, v4 ) ;
```

2.5.4 Double floating-point data type

The double data type allows you to increase the range and precision (or accuracy) of a floating-point number.
The following defines a variable v5 as a double data type:

double类型提高了浮点型数据的取值范围和精度。

```
double v5 ;
```

To display the value in a double variable, either %lf or %f can be used as the format specifier in printf(). For example:

用printf( )函数输出double型变量的值时，既可以使用%lf格式转换符，也可以使用%f格式转换符。

```
printf( "%lf", v5 ) ;
```

2.6 Data type sizes（数据类型的大小）

The next program uses the **sizeof()** operator to display the number of bytes of memory required by some of the common data types in C.

下面的程序使用sizeof( )运算符来显示C语言中一些常用数据类型所占内存空间的字节数。

Program Example P2D

```
1  /* Program Example P2D
2     Program to display the amount of memory required by
3     some of the common data types in C. */
4  #include <stdio.h>
5  int main()
6  {
7   int char_size, int_size, short_size, long_size,
8       float_size, double_size ;
9
10  char_size = sizeof( char ) ;
11  int_size = sizeof( int ) ;
12  short_size = sizeof( short int ) ;
13  long_size = sizeof ( long ) ;
14  float_size = sizeof( float ) ;
15  double_size = sizeof( double ) ;
16
17  printf( " Data type       Number of bytes\n" ) ;
18  printf( " ---------       ---------------\n" ) ;
19  printf( "  char               %d\n", char_size) ;
20  printf( "  int                %d\n", int_size ) ;
21  printf( "  short int          %d\n", short_size ) ;
22  printf( "  long int           %d\n", long_size ) ;
23  printf( "  float              %d\n", float_size) ;
24  printf( "  double             %d\n", double_size ) ;
25  return 0 ;
26 }
```

Lines 7 and 8 of this program define six integer variables.

Lines 10 to 15 use the `sizeof()` operator to assign the size in bytes of each data type to one of the six variables.

Lines 17 to 24 display the value in each of the six variables.

The output from this program is:

```
Data type        Number of bytes
---------        ---------------
  char                1
  int                 4
  short int           2
  long int            4
  float               4
  double              8
```

程序的第7行和第8行定义了6个整型变量，第10 ~15行使用sizeof( )运算符来计算每种数据类型所占内存空间的字节数，然后赋值给相应的6个变量。第17~24行显示这6个变量的值。

Programming pitfalls

1. Do not type a semicolon after either

```
#include <stdio.h>
```

or

```
int main()
```

1. 不要在#include <stdio.h>和int main( )后面添加分号。

2. Comments start with `/*` and end with `*/`.

   Forgetting to end a comment with `*/` may cause part of your program to be ignored by the compiler. For example:

```
printf( "Line1\n" ) ;
/* This comment does not end properly
printf( "Line2\n" ) ;
printf( "Line3\n" ) ;
/* The compiler thinks the comment ends here -> */
printf( "Line4\n" ) ;
```

   The statements above will only display `Line1` and `Line4`. All the statements between `/*` at the start of the second line and `*/` at the end of the fifth line are regarded as a comment.

2. 注释以/*开始，以*/结束。忘记以*/ 结束将导致部分程序被编译器所忽略。

上面的语句只是显示Line1和Line4。位于第2行开头的/*和第5行行末的*/之间的语句都作为注释来处理。

3. Comments cannot be nested, i.e. you cannot have a comment within another comment.

```
/* This is an error because you cannot /*nest*/ comments */
```

3. 注释不能嵌套，即不能在一个注释中包含另一个注释。

4. Use the correct format specifier when displaying a variable using `printf()`. For example,

```
float var ;
var = 123.56 ;
printf( "%d", var ) ;
```

   will incorrectly display the value in the variable `var`, because `var` is a `float` and `%d` is used for displaying an integer.

4. 使用printf( )函数显示变量值时要使用与变量类型相对应的、正确的格式转换符。下面语句就不能正确显示变量var的值，因为var是float类型，而%d是显示整型数据的格式转换符。

5. The second character in the `printf()` format specifiers `%ld` and `%lf` is the letter *ell*, not the number one.

5. printf( )函数的格式转换符%ld和%lf中的第2个字符是字母l，不是数字1。

Quick syntax reference

At the end of each chapter the most important features of the C syntax covered in the text are briefly summarised. While not covering the strict definition of the syntax, which can be complex for a beginner, it should prove to be a useful “memory jog” while writing programs.

在每章的最后，将本章中出现的最重要的C语言语法做一个简要的总结，没包含令初学者头疼的语法的严格定义。在读者编写程序时，这对于唤醒你的记忆是非常有帮助的。

	Syntax	Examples
Start of program	#include <stdio.h> int main() {	
Defining variables	char variable(s) ; int variable(s) ; float variable(s) ; short int variable(s) ; double variable(s) ; unsigned variable(s) ;	char any_letter, y_or_n ; int distance ; float average, pay, tax ; short int temperature ; double total, number ; unsigned int employee_num ;
Assignment	=	tax = 59.75 ;
Comments	/* */	/* Explanatory text. */
Display on the screen	printf(text) ; printf("Tax Program\n") ;	printf(format,variables) ; printf("Tax is %f",tax) ;
End of program	return 0 ; }	

Exercises

1. Which of the following are valid variable names? If valid, do you think the name is a good mnemonic (i.e. reminds you of its purpose)?
   (a) stock_code
   (b) money$
   (c) Jan_Sales
   (d) X-RAY
   (e) int
   (f) xyz
   (g) 1a
   (h) invoice_total
   (i) john's_exam_mark
   (j) default
2. Identify the data type of each of the following constants:
   (a) 'x'
   (b) -39
   (c) 39.99
   (d) -39.0
3. Which of the following are valid variable definitions?
   (a) integer account_code ;
   (b) float balance ;
   (c) decimal total ;
   (d) int age ;
   (e) double int ;
   (f) char c ;

4. Write a variable definition for each of the following:
   (a) integer variables `number_of_transactions` and `age_in_years`
   (b) floating-point variables `total_pay`, `tax_payment`, `distance` and `average`
   (c) a character variable `account_type`
   (d) a double variable `gross_pay`
5. Write the most appropriate variable definition for each of the following:
   (a) the number of students in a class
   (b) an average price
   (c) the number of days since the 1st of January 1900
   (d) a percentage interest rate
   (e) the most common letter on this page
   (f) the population of China (estimated to be 1,339,724,852 in November 2010).
6. Assuming the following:

```
int i ;
char c ;
```

   which of the following are valid C statements?

```
c = 'A' ;
i = "1" ;
i = 1 ;
c = "A" ;
c = '1';
```

7. Write a C program to assign values to the variables in question 4 and display the value of each variable on a separate line.
8. Write a C program that displays the following:

```
***************
* Hello World *
***************
```

9. Write a C program to display your name and address on separate lines.
10. ASCII codes are used to represent letters, digits and other symbols inside the computer's memory. Using the ASCII table in appendix C, look up the ASCII code for each of the following characters:

    `'A' 'B' 'Y' 'Z' 'a' 'b' 'y' 'z' '0' '1' ',' ' '` (a space)

11. In program P2C, change the `%d` in line 14 to `%c` and the `%c` in line 16 to `%d`.
    Compile and run the modified program. Can you explain the output?
    (Hint: See ASCII table in appendix C.)

Chapter Three
Simple Arithmetic Operations and Expressions
第 3 章　简单的算术运算和表达式

3.1　C operators（C 运算符）

Operators are used with one or more *operands* to perform some action in C. Operators are the verbs of the C language. There are a variety of operators in C, as shown in Table 3.1 below.

C**运算符**是指这样的一些动词，它们用于执行C语言中的某些操作，可以带有一个或多个**操作数**。

Table 3.1　operators in C

Operator	Used for
Assignment	assigning values to variables
Arithmetic	basic arithmetic
Relational	comparing values
Logical	combining simple comparisons into more complex tests
Increment	incrementing the value of a variable
Decrement	decrementing the value of a variable
Combined	combining arithmetic and assignment operators
Special	operations on advanced data types, such as pointers (see Chapter Eight)

3.1.1　The assignment operator

The assignment operator (=) is used to assign values to variables. For example, the statement

赋值运算符（=）用于对变量进行赋值操作。

```
v = 1 ;
```

assigns the value `1` to the variable `v`.

Further examples are:

```
year = 1993 ;
value = 100.12 ;
reply = 'y' ;
v1 = v2 = v3 = 12 ;
```

In the last example, the value `12` is assigned to the three variables `v1`, `v2`, and `v3`.

The left-hand side of an assignment will always be a variable. Statements such as:

其中的最后一个语句将12赋值给3个变量v1、v2和v3。赋值运算符的左边必须是变量。

```
2 = v ;    or    2 = 3 ;
```

do not make sense, as `2` is a constant and cannot change its value.

这种赋值没有意义，因为2是一个常量，其值是不能修改的。

3.1.2 Arithmetic operators

There are five arithmetic operators in C, as shown in Table 3.2 below.

C语言中有5种算术运算符。

Table 3.2 arithmetic operators

Operator	Used for
+	addition
–	subtraction
*	multiplication
/	division
%	modulus (this gives the remainder after division)

These arithmetic operations will already be familiar to you. However, while in mathematics you use x for multiplication, in C you use *. The * is used rather than × to avoid any possible confusion with a variable called x. Note also that /, rather than ÷, is used for division.

The next program demonstrates the use of the arithmetic operators.

我们对加法、减法、除法运算符已经非常熟悉了，但是在数学中是使用×进行乘法运算的，而在C语言中则使用*进行乘法运算，这是为了避免运算符×有可能与变量x相混淆。同样注意除法运算使用/ 而非÷。

Program Example P3A

```
1  /* Program Example P3A
2     Introduction to the arithmetic operators in C. */
3  #include <stdio.h>
4  int main()
5  {
6    /* Define the variables used in the program. */
7    int var1, var2 ;
8    /* Assign values to the variables and display
9       the values in the variables. */
10   var1 = 0 ;
11   var2 = 10 ;
12   printf( "var1 is %d and var2 is %d\n", var1, var2 ) ;
13   /* Do some arithmetic with the variables and display
14      the values in the variables. */
15   var2 = var1 + 18 ;
16   printf( "var2 is now %d\n", var2 ) ;
17   var1 = var2 * 3 ;
18   printf( "var1 is now %d\n", var1 ) ;
19   var1 = var2 / 3 ;
20   printf( "var1 is now %d\n", var1 ) ;
21   var2 = var1 - 1 ;
22   printf( "var2 is now %d\n", var2 ) ;
23   var1 = var2 % 3 ;
24   printf( "var1 is now %d\n", var1 ) ;
25   var1 = var1 + 1 ;
26   printf( "var1 is finally %d\n", var1 ) ;
27   var2 = var2 * 5 ;
28   printf( "and var2 is finally %d\n", var2 ) ;
29   return 0 ;
30 }
```

When you run this program, the following will be displayed on your screen:

```
var1 is 0 and var2 is 10
var2 is now 18
var1 is now 54
var1 is now 6
var2 is now 5
var1 is now 2
var1 is finally 3
and var2 is finally 25
```

We will trace the execution of this program by noting the values of the program variables `var1` and `var2` at each line of the program. This is an elementary program trace, and although it may seem very laborious, it is an essential *programming* skill. A program trace simulates the workings of a program and helps in the understanding and *debugging* (removal of errors) of a program.

下面我们通过在程序的每一行标记变量var1、var2的值来跟踪程序的执行过程，虽然这比较费力，但这是一种最基本的程序跟踪方法，也是基本的**编程技巧**。程序跟踪模拟程序的运行过程，有助于理解和调试程序。**调试**就是排除程序中的错误。

Line 7 of this program defines two integer variables, `var1` and `var2`. At this stage we do not know what values are in the memory locations assigned to `var1` and `var2`.

程序的第7行定义了两个整型变量var1和var2。此时，我们并不知道内存单元中的哪些值赋值给了变量var1和var2。

?	?
var1	var2

Line 10 assigns `0` to `var1`, and line 11 assigns `10` to `var2`.

第10行为变量var1赋值0，第11行为变量var2赋值10。

0	10
var1	var2

Line 15 takes the value in `var1` (which is 0) and adds 18 to it, and the result (18) is then assigned to `var2`.

第15行读取变量var1的值（这里为0），加上18后，将结果18赋值给变量var2。

0	18
var1	var2

In line 17, `var2` has a value of 18, which is multiplied by `3`, giving 54, which is assigned to `var1`.

第17行，将变量var2的值18乘以3后，得到结果54，并将其赋值给变量var1。

54	18
var1	var2

Line 19 takes the value in `var2` (18 at this stage) and divides it by `3` and assigns the result (6) to `var1`.

第19行将变量var2的值18除以3后，将结果6赋值给变量var1。

6	18
var1	var2

Line 21 subtracts `1` from the value in `var1` and places the result (5) into `var2`.

第21行从变量var1的值中减去1后，将结果5赋值给变量var2。

6	5
var1	var2

Line 23 demonstrates the modulus operator `%`. In this line `var2` is divided by `3`, and the remainder (2) is assigned to `var1`.

第23行使用取余运算符%，将变量var2对3求余后的结果2 赋值给变量var1。

2	5
var1	var2

Line 25 may seem unusual in that the same variable appears on both sides of the equals sign. This statement does not make sense in algebra. In C, however, it means add 1 to the value in var1 (giving 3), which is then assigned to var1, overwriting its original value of 2.

第25行看上去有些特别，因为相同的变量出现在了等号的两端，这条语句在代数中是没有意义的，但是在C语言中意味着对变量var1的值加1后（得到3），再将结果重新赋值给变量var1，新值3将覆盖变量var1 原来的值2。

3	5
var1	var2

In line 27, the value of var2 is multiplied by 5, giving 25, which is then assigned to var2.

第27行将变量var2的值乘以5后，再将结果25重新赋值给变量var2。

3	25
var1	var2

3.1.3 Increment and decrement operators

It is very common in programming to add or subtract 1 from a variable; so common, in fact, that C provides operators specifically to do these tasks.

In line 25 of program P3A, 1 is added to the value of var1. Instead of writing

事实上，在程序编写过程中，经常要用到对变量进行加1或减1的运算，因此，C语言专门提供了执行这种功能的运算符。

程序P3A的第25行语句对变量var1执行加1操作。

```
var1 = var1 + 1 ;
```

you can write

```
var1++ ;
```

The increment operator ++ adds 1 to the value of a variable. If you wanted to subtract 1 from var1 you could write the statement

自增运算符++是将变量本身的值加1。若要将变量本身的值减1，则可以使用下列语句。

```
var1 = var1 - 1 ;
```

Alternatively, you can use the decrement operator -- and write

也可以使用自减运算符--，即写成下面的语句。

```
var1-- ;
```

The next program demonstrates the increment and decrement operators.

Program Example P3B

```
1  /* Program Example P3B
2     Demonstration of the increment and decrement operators. */
3  #include <stdio.h>
4  int main()
5  {
6    /* Define two variables and initialise them. */
7    int var1 = 1, var2 = 2 ;
8    printf( "Initial values:\n" ) ;
9    printf( "var1 is %d and var2 is %d\n", var1, var2 ) ;
10   var1++ ; /* Increment the value in var1. */
11   var2-- ; /* Decrement the value in var2. */
12   printf( "Final values:\n" ) ;
```

```
13   printf( "var1 is %d and var2 is %d\n", var1, var2 ) ;
14   return 0 ;
15 }
```

The output from this program is:

```
Initial values:
var1 is 1 and var2 is 2
Final values:
var1 is 2 and var2 is 1
```

Line 7 of this program defines the variables `var1` and `var2` as integers and initialises them to `1` and `2`, respectively.
Line 10 adds `1` to `var1` and line `11` subtracts `1` from `var2`.
Line 13 displays the final value of `var1` and `var2`.
The increment operator ++ has two forms: the *pre* and the *post* forms. With pre ++, 1 is added to a variable before that variable is used. With post ++, 1 is added to a variable after that variable is used. The next program demonstrates both of these forms.

程序第7行将变量var1和var2定义为整型，并将其分别初始化为1和2。
程序第10行将变量var1的值加1，第11行将变量var2的值减1。
第13行显示变量var1和var2的最终的值。
自增运算符++有两种形式，一种为**前缀**形式，一种为**后缀**形式。前缀的自增运算符是在变量使用前对变量的值增1。而后缀的自增运算符是在变量使用后对变量的值增1。

Program Example P3C

```
1  /* Program Example P3C
2     Demonstration of prefix and postfix ++ operators. */
3  #include <stdio.h>
4  int main()
5  {
6    int var1 = 11, var2 = 22, var3, var4 ;
7
8    var3 = ++var1; /* prefix : var1 is 12, var3 is 12 */
9    var4 = var2++ ; /* postfix: var4 is 22, var2 is 23 */
10   printf( "var1 is %d, var3 is %d\n", var1, var3 ) ;
11   printf( "var4 is %d, var2 is %d\n", var4, var2 ) ;
12   return 0 ;
13 }
```

The output from this program is:

```
var1 is 12, var3 is 12
var4 is 22, var2 is 23
```

The program starts by initialising the variables `var1` to `11` and the variable `var2` to `22`.
In line 8 the variable `var1` is incremented first and then its new value (=12) is assigned to `var3`. This is an example of a *prefix* operation.
In contrast, line 9 is an example of a *postfix* operation. In line 9, `var4` is assigned the value of `var2` (=22) and then the value of `var2` is incremented to 23.

在程序开头，变量var1初始化为11，变量var2初始化为22。
程序第8行语句先将变量var1的值增1，然后将增1后的新值12赋值给变量var3。这是一个**前缀操作**的例子。
相反，第9行是一个**后缀操作**的例子，第9行语句先将变量var2的值22赋值给变量var4，然后将变量var2的值增1变成23。

If the ++ is before a variable, the variable is incremented before it is used. If the ++ is after the variable, the variable is used and then incremented. The difference between prefix and postfix is only relevant where an assignment is involved. So in line 10 of program P3B, it doesn't matter whether you use var1++ or ++var1.

如果++写在变量的前面，那么在变量使用之前，先对其执行加1操作，而如果++写在变量的后面，那么就先使用变量的当前值，然后再对其进行加1操作。其前缀形式与后缀形式的差异仅在赋值操作中才能体现出来。例如，程序P3B的第10行语句使用var1++还是++var1是无关紧要的。

The decrement operator -- also has prefix and postfix forms. If the -- is before a variable, the variable is decremented before it is used. If -- is after the variable, the variable is used and then decremented. Again this is only relevant if an assignment is involved. In line 11 of program P3B you could use either var2-- or --var2.

自减运算符--同样也有前缀和后缀形式。如果--写在变量的前面，那么在变量使用之前，先对其执行减1操作，而如果--写在变量的后面，那么就先使用变量的当前值，然后再对其进行减1操作。同样地，自减运算符的前缀形式与后缀形式的差异也是仅在赋值操作中才能体现出来。例如，程序P3B的第11行语句既可以使用var2--，也可以使用--var2。

3.1.4 Combined operators

The ++ operator is very convenient for replacing a statement such as

```
var = var + 1 ;
```

with

```
var++ ;
```

or

```
++var ;
```

C allows you to take this convenience one step further with the combined operators. Instead of writing

C语言还允许使用复合运算符来简化语句。

```
var = var + 3 ;
```

which adds 3 to the value of var, you can write

对变量var自身执行加3操作，也可以将其写为如下形式。

```
var += 3 ;
```

The += adds the value on its right to the variable on its left. There are five combined arithmetic operators: +=, -=, *=, /=, and %=, corresponding to the five arithmetic operators +, -, *, /, and %. (See Table 3.3.)

运算符+=执行的操作是，将其左侧变量的值加上右侧的数值后，再重新赋值给左侧的变量。C 语言中有5 种复合的算术运算符：+=，-=，*=，/=，%=，分别对应+、-、*、/、%这5 种算术运算。

Table 3.3 examples of combined operators

Operator	Examples	Equivalent
+=	count += 11 ;	count = count + 11 ;
	a += b ;	a = a + b ;
-=	count -= 20 ;	count = count - 20 ;
	a -= b ;	a = a - b ;
*=	money *= 2 ;	money = money * 2 ;
	a *= b ;	a = a * b ;

（cont.）

Operator	Examples	Equivalent
`/=`	`money /= 2 ;` `a /= b ;`	`money = money / 2 ;` `a = a / b;`
`%=`	`pence %= 100 ;` `a %= b ;`	`pence = pence % 100 ;` `a = a % b ;`

Although the C language does not require you to use any of the combined operators, it is well worth while learning how to use them, as you will come across them quite frequently in C programs.

虽然C语言并不要求一定要使用复合的运算符，但是由于它们在C程序中频繁出现，因此还是值得好好学习如何使用它们。

3.2 Operator precedence（运算符优先级）

Consider the following statement:

```
var = 2 + 7 * 8 ;
```

Does this mean

(a) that `2` is added to `7`, giving 9, which is multiplied by `8`, giving `var` a value of 72,

or

(b) that `7` is multiplied by `8`, giving 56, which is added to `2`, giving `var` a value of 58?

下面的语句究竟是(a)先执行“2+7”得9，然后乘以8得72，赋值给变量var；还是(b)先执行“7*8”得56，然后加2得58，赋值给变量var呢？

Clearly the order of evaluation is important. With (a) you get 72 and with (b) you get 58.

显然，求值的顺序非常重要，按(a)计算得72，而按(b)计算则得58。

There must be some rules to remove any ambiguity present in a C statement such as the one above. *Operator precedence*, as shown in Table 3.4, provides these rules.

必须有一些规则来避免如前所述的C语句的歧义性。**运算符优先级**就提供了这样的规则。

Table 3.4 operator precedence

Operator	Precedence	Meaning	Associativity
`-`	Highest	Unary minus	Right to left

`*`
`/`
`%`
	Lower	Multiplication, division, and modulus	Left to right

`+` `-`	Lowest	Addition and subtraction	Left to right

From this table, multiplication and division have a higher priority than addition and subtraction, so these operators get evaluated first. Thus the statement

乘法和除法的优先级高于加法和减法的优先级，所以先执行乘法和除法。

```
var = 2 + 7 * 8 ;
```

evaluates to 2 + 56 = 58.

You can use parentheses to change the order of evaluation. Thus, the statement

```
var = ( 2 + 7 ) * 8 ;
```

evaluates to 9 * 8 = 72, because any expressions contained within parentheses get evaluated first.

可以使用括号改变运算符的执行顺序，因为括号内的表达式先被计算，因此下面语句的计算结果为9*8 = 72。

Expressions containing operators of the same precedence are evaluated according to their *associativity*, as shown in Table 3.4. Associativity gives the order in which operators of the same precedence are evaluated. For example, in the statement

当表达式中含有相同优先级的运算符时，表达式的求值顺序取决于如表3.4所示的运算符的**结合性**。结合性决定了相同优先级的运算符的执行顺序。

```
var = 1 + 6 * 9 % 5 / 2 ;
```

the *, / and % have equal precedence, so which is done first? The *, / and % associate left to right, so the order is:

其中，*、/ 和%具有相同的优先级，其求值顺序如何呢？由于*、/ 和%都是左结合的，因此其求值顺序如下。

```
1 + 6 * 9 % 5 / 2 = 1 + 54 % 5 / 2
  = 1 + 4 / 2 = 1 + 2 = 3
```

Precedence and associativity are a source of confusion for C novices. I suggest that you use parentheses to group the variables, constants and operators in a way that is clear to you. Expressions within parentheses are evaluated first; therefore, using parentheses will remove any confusion about which operations are done first. For example, the last statement could be written much more clearly as:

运算符的优先级和结合性常使C语言的初学者感到困惑，建议读者在编程过程中，按照表达式的执行顺序，使用括号将语句中的变量、常量和运算符清晰地分组。由于带括号的表达式最先被计算，因此用括号可以清晰地说明表达式的顺序，避免歧义。

```
var = 1 + ( ( 6 * 9 ) % 5 ) / 2 ;
```

and the result will be the same.

The unary minus operator, like the binary minus operator, is represented as -.

一元的负号运算符，与二元的减法运算符类似，记为-。

In the statement

```
var = -3*2-1 ;
```

the first - is a unary minus and the second - is a binary minus. The unary minus appears before an operand and the binary minus appears between two operands. Because the unary minus has highest priority, the above statement is equivalent to

第一个“-”是一元的负号运算符，而第二个“-”则是二元的减法运算符。一元的负号运算符在单个操作数前出现，而二元的减法运算符则出现在两个操作数之间。

```
var = ((-3)*2)-1 ;
```

and not

```
var = -(3*2-1) ;
```

In the first case `var` is assigned a value of −7, while in the second case −5 is assigned to `var`.

3.3 Type conversions and casts（类型转换与强制类型转换）

Consider the following program, which divides an integer variable `var1` by a floating-point variable `var2`, placing the result in a floating-point variable `var3`.

下面的程序用一个整型变量var1除以一个浮点型变量var2，并将其结果值赋值给浮点型变量var3。

Program Example P3D

```
1  /* Program Example P3D
2     Demonstration of mixed data type expressions. */
3  #include <stdio.h>
4  int main()
5  {
6    int var1 = 10 ;
7    float var2 = 2.4 ;
8    float var3 ;
9    int var4 ;
10   float var5 ;
11   var3 = var1 / var2 ; /* Mixed expression assigned to a float    */
12   var4 = var1 / var2 ; /* Mixed expression assigned to an int     */
13   var5 = var1 / 4 ;    /* Non-mixed expression assigned to a float */
14   printf( "var3 = %f var4 = %d var5 = %f\n", var3, var4, var5 ) ;
15   return 0 ;
16 }
```

The output from this program is:

```
var3 = 4.166667 var4 = 4 var5 = 2.000000
```

Line 11 is an example of a *mixed expression*, where the variable `var1` of type `int` is divided by the variable `var2` of type `float`. The value of the variable `var1` is automatically converted to a `float` before being divided by the value of `var2`. The result of the division is a value of type `float`, which is then assigned to the `float` variable `var3`. (Note that only a copy of the value of `var1` is converted to a `float`. The variable `var1` is still an `int`.)

第11行的语句中存在一个**混合表达式**，变量var1是一个int类型变量，被一个float类型变量var2去除，变量var1在被var2除之前由系统自动转换为float类型变量，因此所得结果为float类型，该结果赋值给float类型变量var3（注意变量var1本身的类型仍为int类型，只是该变量的副本转换为float 类型）。

In line 12, the result of the division is assigned to an `int` type variable `var4`, resulting in a loss of the fractional part of the number.

在第12行，除法的结果赋值给int类型变量var4，因此结果的小数部分丢失了。

In line 13, the division involves two integers, so there is no need for any type conversions. The result of the division is the integer value 2, which is converted to a `float` and assigned to `var5`.

在第13行，由于除数和被除数均为整数，因此不需要进行类型转换，除法结果为整数值2，将其赋给float类型变量var5时，系统自动将其转换为float类型。

When doing calculations involving mixed data types, C ranks the data types in this order:

当进行混合类型数据间的运算时，C语言对数据类型优先级的排序如下所示。

```
char < short < int <long int <float < double
```

and a signed data type is less than the corresponding unsigned data type (for example, a `signed int < unsigned int`). When doing calculations involving mixed data types, C automatically converts the value in the lower data type to a higher type. This is called a *promotion* or *widening* of the data. Promotion will cause no loss of data, because the higher data types occupy more memory than the lower types and can therefore hold the data precisely. On the other hand, when data is assigned to a variable of a lower type, *demotion* or *narrowing* occurs. Demotion may result in a loss of data, because the lower data type may not have enough bytes of memory to store the higher data type.

In addition to automatic conversion, C allows you to perform manual conversion with casts. The general format of a cast is:

有符号数据类型的优先级小于无符号数据类型的优先级。

当在混合类型数据间进行运算时，C语言自动将优先级较低的数据类型转换为优先级较高的数据类型，这称为**类型提升**或**类型扩展**。由于高级别的数据类型占用的内存空间比低级别的数据类型要多，可以保持数据的精度，因此数据提升可以避免数据丢失情况的发生。另一方面，如果将数据赋值给比其数据类型级别低的变量，那么就会发生**类型降级**或**类型缩小**，由于低级别的数据类型占用的字节数少于高级别的数据类型，因此，数据降级将导致数据丢失现象的发生。

除了自动类型转换，C语言还允许用户使用强制类型转换运算符进行手动类型转换。

```
(type)expression
```

Using a cast is equivalent to assigning the expression to a variable of the specified type and using that variable in place of the expression. For example, if you change line 13 of program P3D to:

使用强制类型转换运算符相当于将一个表达式赋值给特定类型的变量，然后用该变量替代该表达式。例如，如果将程序P3D的第13行改为如下形式。

```
var5 = (float)var1 / 4 ;
```

the value of `var1` is cast from an integer to a floating-point value. The expression is now mixed, and the integer value 4 will be promoted to a `float`. The result of the division is now 2.5, which is assigned to `var5`. Without the cast, `var5` would have a value of 2.0. You can also get the same result by rewriting line 13 as:

那么就将整型变量var1的值强制转换为了浮点类型，于是表达式(float)var1/4就变成了一个混合类型的表达式，整型常量值4被类型提升为float类型，因此除法的结果为2.5，将其赋值给float类型的变量var5，而如果不进行强制类型转换，那么变量var5的值将为2.0。将第13行语句改为如下形式将得到同样的结果。

```
var5 = var1 / 4.0 ;
```

Here `4.0` is a floating-point constant. The expression `var1 / 4.0` is now mixed, and `var1` is therefore promoted to a `float` automatically.

因为4.0是float类型的常量，表达式var1/4.0为混合类型的表达式，因此变量var1的类型将自动提升为float类型。

Programming pitfalls

1. A typing error may result in a statement that does nothing but that is valid nonetheless.
   For example,

```
a + b ;   /* Does nothing. */
```

   This was probably meant to be:

```
a += b ;  /* Adds b to a.    */
```

   or

```
a = b ;   /* Assigns b to a. */
```

1. 键入错误有可能导致语句什么也不做，不过它仍然是合法的。

2. You must initialise a variable before using it in an arithmetic expression.

```
int counter ;  /* What value is in counter? */
counter++ ;
```

   This code adds 1 to the value in the memory location occupied by `counter`. We have no idea what this value is, as the program did not assign `counter` a value. The problem is fixed by initialising `counter` to some value.
   For example:

```
int counter = 0 ;
```

2. 在算术表达式中，使用变量前必须先对其进行初始化。

   这段代码是对变量counter占据的内存单元中的值执行加1 操作，但由于并没有为counter进行赋值，因此不知道counter占据的内存单元中的值是什么，其解决方法是在使用counter前对其进行初始化，为它赋一个值。

3. Be aware of the operator precedence rules. If you are unsure, use parentheses. If you are using parentheses in an expression, count the number of opening and closing parentheses. They should be equal.

3. 注意运算符的优先级顺序，如果对此不确定，那么可以在表达式中使用圆括号来确定其执行顺序。表达式中的左括号“(”数量应该与右括号“)”数量相等。

4. Each variable has an associated data type (`int`, `float`, etc.). Be careful not to go outside the range of valid values for a variable. For example, a `short int` cannot hold values bigger than 32,767 or smaller than -32,768.

4. 每个变量都有一个相关的数据类型（如int、float等）。注意变量的值不要超出其类型所确定的合法取值范围。

5. You may not always get what you expect when doing arithmetic in C. For example:

```
int a = 100 ;
int b = 8 ;
float f ;

f = a / b ;
```

   The variable f will contain 12, and not the 12.5 you would expect. As `a` and `b` are integers, integer division is performed, resulting in the loss of the fractional part of the result. Use a cast if the fractional part of the result is required. For example:

```
f = (float)a / b ;
```

5. 在执行C算术运算中，有可能得不到预期的结果。

在该段程序中，变量f将得到值12，而不是预期的12.5，这是由于a和b为整型变量，因此程序将执行整型除法，从而丢失了结果的小数部分。如果要保留结果的小数部分，那么需要进行强制类型转换。

Quick syntax reference

	Syntax	Examples

Arithmetic operators	+
	 –
	 *
	 /
	 %
		`vat = sales*0.21 ;`
		 `remainder = number%10 ;`
		 `average = (n1+n2)/2 ;`

Increment and decrement	++
	 --
		`n1++ ;`
		 `n2-- ;`
		 `n3 = ++n1 ;`
		 `n4 = n2-- ;`

Combined operators	+= –= *= /= %=	`n1 += 10 ;` `n2 *= n1 ;`
Casting	(type)	`answer = (float)n1/n2 ;`

Exercises

1. Convert the following mathematical equations into valid C statements:

   (a) $m = \frac{y_1 - y_2}{x_1 - x_2}$

   (b) $y = mx + c$

   (c) $a = \frac{b}{c} - \frac{d}{e}$

   (d) $c = \frac{5(F-32)}{9}$

   (e) $s = ut + \frac{1}{2}at^2$

2. Assuming the following variable definitions:

```
int a = 1, b = 10, c = 5 ;
int d ;
```

   what is the value of `d` after each of the following statements?

   (a) `d = b / c + 1 ;`

   (b) `d = b % 3 ;`

   (c) `d = b - 3 * c / 5 ;`

   (d) `d = b * 10 + c - a * 5 ;`

   (e) `d = ( a + b - 1 ) / c ;`

   (f) `d = ( ( -a % c ) + b ) * c ;`

   (g) `d = --a ;`

3. Assuming the same variable definitions as in question 2, correct the errors in the following C statements:

   (a) `d = 2(b + C) ;`

   (b) `d = 5b + 9c ;`

   (c) `d = b - 3 X 19 ;`

   (d) `d = b.c + 10 ;`

   (e) `d = ( a + b ) / c ;`

4. Write suitable statements to perform the following:

   (a) add `1` to `num1`, placing the result in `num1`

   (b) add `2` to `num1`, placing the result in `num2`

   (c) add `2` to `num2`, placing the result in `num2`

   (d) subtract `1` from `num1`, placing the result in `num1`

   (e) subtract `2` from `num2`, placing the result in `num2`

5. Assuming the following definitions:

```
int a = 12, b = 0, c = 3 ;
int d ;
```

   what is the value of `a`, `b`, `c` and `d` after each of the following statements?

   (a) `a++ ;`

   (b) `b-- ;`

   (c) `d = ++c ;`

   (d) `d = c-- ;`

   (e) `d = a++ - 2 ;`

   (f) `d = a++ + b++ - c-- ;`

6. Assuming the following definitions:

```
int a = 1, b = 2, c = 3 ;
```

   what is the value of `a`, `b` and `c` after each of the following statements?

   (a) `a += b ;`

   (b) `a /= 3 ;`

   (c) `a *= c ;`

   (d) `a %= 2 ;`

   (e) `a += b+1 ;`

   (f) `a += ++b ;`

7. Assuming the following definitions:

```
char ch_val ; int int_val ; short short_val ;
float float_val ; double double_val ;
unsigned int unsigned_int_val ;
```

   which of the following may lose data because of demotion?

   (a) `short_val = int_val ;`

   (b) `int_val = ch_val ;`

   (c) `double_val = float_val ;`

   (d) `int_val = float_val ;`

   (e) `int_val = unsigned_int_val ;`

8. Assuming the same variable definitions as in question 7, what is the data type of each of the following expressions?

   (a) `int_val * float_val ;`

   (b) `float_val + int_val / 100 ;`

   (c) `ch_val + short_val + int_val ;`

(d) `(double)int_val + double_val + float_val ;`

(e) `(int)float_val * float_val / int_val ;`

(f) `int_val + 3.0 ;`

9. Assuming the following variable definitions:

```
int a = 5, b = 4 ;
float c = 3.0, d ;
```

what is the value of `d` after each of the following?

(a) `d = a / b ;`

(b) `d = (float)a / b ;`

(c) `d = c / b ;`

(d) `d = (int)c / b ;`

(e) `d = a / 2 ;`

(f) `d = a / 2.0 ;`

(g) `d = (float)a / 2 ;`

(h) `d = (int)c % 2 ;`

10. Write a program to compute the volume and the surface area of a box with a height of 10 cm, a length of 11.5 cm and a width of 2.5 cm.

11. Write a program to do the following:

(a) calculate and display the sum of the integers 1 to 5

(b) calculate and display the average of the floating-point numbers 1, 1.1, 1.2 ... 2.0.

Chapter Four
Keyboard Input and Screen Output
第 4 章　键盘输入和屏幕输出

4.1 Simple keyboard input（简单的键盘输入）

Here is a simple program to demonstrate keyboard input and screen output:

下面是演示键盘输入和屏幕输出的一个简单程序。

Program Example P4A

```
1  /* Program Example P4A
2      To demonstrate simple keyboard input and screen output. */
3  #include <stdio.h>
4  int main()
5  {
6    int var ;
7    printf( "Please type a number: " ) ;
8    scanf( "%d", &var ) ;
9    printf( "\nThe number you typed was %d\n", var ) ;
10   return 0 ;
11 }
```

This program simply asks the user for a number and displays the number entered on the screen.

Line 7 displays the message:

```
Please type a number:
```

The **scanf()** function in line 8 causes the computer to wait indefinitely until you type a number and press the Enter key. When you type in a number (e.g. 11) followed by the Enter key, the program continues to line 9 and displays:

```
The number you typed was 11
```

这个程序首先要求用户输入一个数字，然后将其在屏幕上显示出来。

第7行语句的功能是显示如下信息。第8行语句中scanf( )函数的功能是等待用户输入一个数字，然后按回车键，如果用户没有任何输入，那么程序将无限期地等待。当用户输入一个数字（例如11）并按回车键后，程序将继续执行第9行的语句，并且在屏幕上输出如下信息。

The `scanf()` function in line 8 has an ampersand (**&**) before the variable `var`. An ampersand preceding a variable refers to the address of that variable. Therefore, `&var` refers to the address of the variable `var`. The address of a variable is the location in memory where the value of that variable is stored. Just as a letter is delivered to the address specified on the envelope, the `scanf()` function stores the number you typed at the address specified by `&var`.

程序第8行scanf( )函数中的变量var前面有一个取地址运算符（&），它的作用是取出变量在内存中的地址，因此&var 的值实际上是变量var 对应的地址。变量的地址就是变量存储在内存中的存储单元的地址，就像寄信时写在信封上的地址一样，scanf( )函数将用户输入的数值存储到由&var指定的变量var的内存地址中。

Like the `printf()` function, the `scanf()` function has a format string. In line 8 the format string is "`%d`". The letter `d` specifies that the input data is a decimal integer value. Different format strings are used for different types of variables. For example, "`%f`" is used to input a value to a `float`, and "`%lf`" is used to input a value to a `double` variable.

与printf( )函数类似，scanf( )函数中也有一个格式字符串。不同的格式字符串用于不同类型的变量。例如，%d用于输入一个十进制整型数值，%f用于输入一个浮点型数值，%lf用于输入一个双精度型数值。

The next program inputs two floating-point numbers from the keyboard and displays the result of their addition.

下面的程序从键盘输入两个浮点型数值，然后显示它们的和。

Program Example P4B

```
1  /* Program Example P4B
2     To input two numbers and display their sum. */
3  #include <stdio.h>
4  int main()
5  {
6    float sum ;
7    float num1, num2 ;
8    printf( "Type in 2 numbers. Press Enter after each number.\n" );
9    scanf( "%f", &num1 ) ;
10   scanf( "%f", &num2 ) ;
11   sum = num1 + num2 ;
12   printf( "\n%f plus %f is %f\n", num1, num2, sum ) ;
13   return 0 ;
14 }
```

This program reads the two floating-point numbers in lines 9 and 10. Note that both the `scanf()` functions have a `%f` for floating-point numbers. Statements 9 and 10 can be combined into the single statement

这个程序在第9行和第10行读入两个浮点型数值。注意这两个scanf( )函数都用%f来输入浮点型数据。第9行和第10行的语句可以合并为下面这条语句。

```
scanf( "%f%f", &num1, &num2 ) ;
```

When you run this program you will get the following on your screen:

```
Type in two numbers. Press Enter after each
number.
```

If you type **12.25** followed by Enter and then type **1.6** followed by Enter, the program will display:

```
12.250000 plus 1.600000 is 13.850000
```

4.2 Using a width and precision specification in `printf()` [在函数printf( )中使用域宽和精度说明]

When a floating-point number is displayed using `printf()`,it is displayed showing six decimal places. You can tidy up this output by a small change in the format specifier of the `printf()` function.

使用printf( )函数显示浮点数时，显示的数据将保留小数点后的6位小数，可以改变printf( )函数中的格式转换符，使输出更简洁。

If you replace line 12 in the example program P4B with

```
printf( "\n%5.2f plus %5.2f is %6.2f", num1, num2, sum ) ;
```

the output will become:

```
12.25 plus 1.60 is 13.85
```

The %5.2f specifies that num1 and num2 are to be displayed using a total width of five columns and with a precision of two decimal places. Similarly, sum is displayed using a total of six columns and two decimal places.

%5.2f指定了变量num1和num2的输出域宽为5列，数据精度为2位小数，同理变量sum的输出域宽为6列，数据精度为2位小数。

What if the numbers are too big to fit into the width specified? If they are, then the width will be ignored and the full number will be displayed. On the other hand, if there are more than two decimal places in the number, the number will be rounded to two decimal places before being displayed.

如果数据过大，无法满足指定的域宽要求时，那么结果如何呢？这时系统将忽略程序指定的域宽值，将数据完整地显示出来。另一方面，如果数据的小数位大于指定的精度值，那么在显示数据以前，系统将对超出2位小数的部分进行舍入处理。

The next program demonstrates the use of the width and precision specifications in the printf() function.

下面的程序将演示如何在printf( )函数中指定输出数据的域宽和精度。

Program Example P4C

```
1  /* Program Example P4C
2     To demonstrate width and precision specifications
3     in the printf() function.                          */
4  #include <stdio.h>
5  int main()
6  {
7   float num1 = 100.15799, num2 = 12.55, num3 = 1.7 ;
8   int num4 = 621, num5 = 10, num6 = 5 ;
9
10  printf( "printf WITHOUT width or precision specifications:\n" );
11  printf( "%f\n%f\n%f\n", num1, num2, num3 ) ;
12  printf( "%d\n%d\n%d\n\n", num4, num5, num6 ) ;
13  printf( "printf WITH width and precision specifications:\n" ) ;
14  printf( "%5.2f\n%6.1f\n%3.0f\n", num1, num2, num3 ) ;
15  printf( "%5d\n%6d\n%2d\n", num4, num5, num6 ) ;
16  return 0 ;
17 }
```

Note that line 9 is completely blank. Blank lines and spaces are used in a program to improve readability. A blank line is used here to separate the variable definitions from the rest of the program statements.

程序的第9行完全是空行，在程序中增加空行和空格可以增强程序的可读性，这里的空行是为了将变量定义与程序中的其他语句区分开来。

The output from this program is:

```
printf WITHOUT width or precision specifications:
100.157990
12.550000
1.700000
621
```

```
10
5

printf WITH width and precision specifications:
100.16
  12.6
  2
  621
    10
 5
```

4.3 Single-character input and output (单个字符的输入和输出)

The examples we have used so far have involved numeric input and output. You can use `scanf()` and `printf()` with `%c` in the format string to read and display single characters. This is shown in the next program.

到目前为止我们介绍的例子都是数值的输入和输出。scanf( )函数和printf( )函数也可以实现单个字符的输入和输出，此时需要使用%c格式转换符。

Program Example P4D

```
1  /* Program Example P4D
2     To input a single character and display it. */
3  #include <stdio.h>
4  int main()
5  {
6    char ch ;
7
8    printf( "Press a key and then press Enter " ) ;
9    scanf( "%c", &ch ) ;
10   printf( "You pressed the %c key\n", ch ) ;
11   return 0 ;
12 }
```

Running this program will produce the following:

```
Press a key and then press Enter A
You pressed the A key
```

Note: The `%c` conversion character will accept any character from the keyboard, including whitespace characters (keyboard keys such as Tab, Enter, and Space are whitespace characters). To ignore whitespace characters insert a space before `%c` in the format string of the `scanf()` function. C has also two very fast and efficient functions specially designed for single-character input and output. These are **getchar()** and **putchar()**. The next program shows how to use these two functions.

注意：格式转换符%c使程序接收从键盘输入的任何字符，包含一些空白字符，如制表符、回车符和空格等。在scanf( )函数格式字符串中的格式转换符%c的前面插入一个空格，可以使程序忽略这些空白字符。
C语言为单个字符的输入和输出提供了两个非常快速有效的函数，它们是getchar( )和putchar( )。下面的程序介绍如何使用这两个函数。

Program Example P4E

```
1  /* Program Example P4E
2     To demonstrate getchar() and putchar(). */
3  #include <stdio.h>
```

```
4  int main()
5  {
6    char ch;
7
8    printf( "Press a key and then press Enter " ) ;
9    ch = getchar() ;
10   printf( "\nYou pressed " ) ;
11   putchar( ch ) ;     /* Display the character and */
12   putchar ( '\n' ) ; /* skip to a new line.       */
13   return 0 ;
14 }
```

Line 9 reads a single character from the keyboard and places it in the variable `ch`. Line 11 then displays the value of `ch`. Running this program will produce the following:

程序的第9行从键盘读入单个字符，然后将其保存到变量ch中，第11行向屏幕输出变量ch的值，运行该程序将得到如下结果。

```
Press a key and then press Enter A
You pressed A
```

Programming pitfalls

1. Do not omit the address operator & when using scanf(). For example, if num1 is a floating-point variable, then

```
scanf( "%f", num1 ) ;  /* Error! */
```

is invalid and should be:

```
scanf( "%f", &num1 ) ; /* Correct. */
```

1. 使用scanf( )函数时，千万不要漏掉取地址运算符&。

2. Do not place a message in scanf(), as in this statement:

```
scanf( "Input two numbers %f%f", &num1, &num2 ) ;
/* Error! */
```

This will not work. If you want a message displayed, you must use printf(), as in the following:

```
printf( "Input two numbers " ) ;
scanf( "%f%f", &num1, &num2 ) ;
```

2. 不要在scanf( )函数中添加任何信息。

若要在屏幕上输出提示信息，应使用printf( )函数。

3. When using scanf() or printf(), you must have a conversion character for each variable. For example:

```
scanf( "%f%f", &num1, &num2, &num3 ) ; /* Error! */
```

This statement is invalid, because there are three variables and only two conversion characters in the format string.

3. 在使用scanf( )和printf( )函数时，每个变量必须都有一个与其相对应的格式转换符。下列语句是错误的，因为在其输出列表中有3个变量，但格式字符串中只有两个格式转换符。

4. Use the correct conversion character for the type of variable you are using in scanf(). For example, if num4 is a floating-point variable, then

```
scanf( "%d", &num4 ) ; /* Error! */
```

is invalid and should be:

```
scanf( "%f", &num4 ) ; /* Correct. */
```

4. 使用scanf( )函数时，必须根据变量的类型使用正确的格式转换符。

5. Different data types require different format specifiers in scanf(). For example, to read a value into a double variable, "%lf" is used. See Appendix D.

5. 若用scanf( )函数输入不同类型的数据，需使用不同的格式转换符。例如，用scanf( )输入双精度实型变量的值，应使用%lf格式转换符。

Quick syntax reference

	Syntax	Examples

Input data from the keyboard	`scanf(format,&variables) ;`	`char grade ;`
		`int age ;`
		`scanf("%c %d",&grade,&age) ;`

Input a single character from the keyboard	`variable=getchar() ;`	`char char_in ;` `char_in=getchar() ;`
Output a single character to the screen	`putchar(variable) ;`	`char char_out = 'x';` `putchar(char_out) ;`

Exercises

1. What is wrong with this program?

```
#include <stdio.h>
int main()
{
  int num;
  printf( "Please type a number followed by Enter" ) ;
  scanf( "%f", num ) ;
  printf( "The number you typed was: %d", num ) ;
  return 0 ;
}
```

2. Write a single `scanf()` statement to input values from the keyboard for each of the following:

   (a) `int first ;`

   (b) `int second, third, fourth ;`

   (c) `float principal, rate, time ;`

   (d) `char keyval1, keyval2 ;`

   (e)
   ```
   char c ;
   int i ;
   float f ;
   double d ;
   ```

3. Write a program to input four numbers and display them in the reverse order in which they were input.
4. Suppose that `v1`, `v2` and `v3` are three floating-point variables with values 5.0, -4.5, and 11.25, espectively. Write a `printf()` statement to display this line:

```
v1 = 5    v2 = -4.5    v3 = 11.25
```

5. Assuming the human heart rate is seventy-five beats per minute, write a program to ask a user their age in years and to calculate the number of beats their heart has made so far in their life.
6. Write a program to accept a temperature in degrees Fahrenheit and convert it to degrees Celsius. Your program should display the following prompt:

```
Enter a temperature in degrees Fahrenheit:
```

You will then enter a decimal number followed by the Enter key.
The program will then convert the temperature by using the formula

Celsius = (Fahrenheit -32.0) * (5.0 / 9.0)

Your program should then display the temperature in degrees Celsius using an appropriate message.

7. Make changes to the program developed in exercise 6 to accept the temperature in degrees Celsius and convert it to degrees Fahrenheit.
（Fahrenheit：华氏温度；Celsius：摄氏温度）
8. Write a program to input three floating-point numbers from the keyboard and to calculate
   (a) their sum and
   (b) their average.
   Display the results to three decimal places.

9. Write a program to input two integers from the keyboard and display the first number as a percentage of the second number.

   Display the percentage value with one decimal place.

   For example, assuming that the numbers input are **5** and **40**, your output should look like this:

```
5 is 12.5 percent of 40
```

10. Write a program to input two integers from the keyboard and to display the result of dividing the second number into the first.

    For example, if you input 123 and 12, display the results in the following format:

```
     10 Remainder = 3
   ----
12 )123
```

(Hint: use the modulus operator % to get the remainder 3, and use integer division to get the quotient 123.)
（modulus operator：求余运算符）

Chapter Five
Control Statements: if and switch
第 5 章　控制语句： if 和 switch

All the programs we have written so far execute one statement after the other, starting at the first statement and finishing at the last. Only the simplest of problems can be solved using this sequential top-to-bottom approach. Control statements are used to modify this sequential execution of program statements. This chapter covers the if and the switch control statements.

到目前为止我们编写的程序都是按顺序执行的，程序从第一条语句开始，逐条执行到最后一条语句结束，只有最简单的问题才可以用这种自上而下的顺序结构来解决。为了改变程序语句的执行顺序，可以使用C语言提供的控制语句。本章将介绍if和switch这两种控制语句。

5.1 The if statement（if 语句）

The **if** statement starts with the keyword if followed by an expression in parentheses. If the expression is found to be true, then the statement following the if is executed. If the expression is untrue, then the statement following the if is not executed. For example:

if语句是由关键字if与其后括号中的表达式组成的，如果该表达式的值为真，那么紧随if其后的语句被执行，如果该表达式的值为假，那么紧随if其后的语句不会执行。

```
if ( account_balance < 0 )
   printf( "Your account is in the red\n" ) ;
```

This statement tests whether the variable account_balance is less than 0. If it is less than 0, then the printf() is executed and the message is displayed. If account_balance contains a number greater than or equal to 0, then the printf() is not executed. The < is called a *relational operator*. The full list of relational operators is given in Table 5.1.

这条语句测试变量account_ balance的值是否小于0，如果它小于0，那么就执行printf( )，在屏幕上输出信息“Your account is in the red”，如果变量account_balance的值大于或等于0，那么printf( )就不会执行。其中，<是一个**关系运算符**。表5.1给出了所有的关系运算符及其含义。

Table 5.1　relational operators

Operator	Meaning
==	equivalent to
!=	not equal to
<	less than
>	greater than
<=	less than or equal to
>=	greater than or equal to

It is important to note that when you are testing for equality the operator used is == and not =.
For example:

需要特别注意的是，比较相等时，应使用== ，而不能使用=。

```
if ( account_balance == 0 )
   printf( "Your account has no money in it\n" ) ;
```

Program Example P5A

```
1  /* Program Example P5A
2     Demonstration of the if statement. */
3  #include <stdio.h>
4  int main()
5  {
6    float account_balance ;
7
8    printf( "What is your account balance? " ) ;
9    scanf( "%f", &account_balance ) ;
10   if ( account_balance < 0 )
11     printf( "Your account is in the red\n" ) ;
12   if ( account_balance >= 0 )
13     printf( "Your account is in the black\n") ;
14   return 0 ;
15 }
```

When you run this program it will ask you for your account balance. Type in your account balance and press Enter. If your account balance is negative, the statement on line 11 is executed; if it is greater than or equal to 0, then line 13 is executed. Here is a sample run of this program:

运行这个程序时，首先要求用户输入账户余额，当用户键入账户余额数值并按回车键后，若账户余额为负值，则执行第11行的语句，如果账户余额大于等于0，则执行第13行的语句。下面是程序运行的简单示例。

```
What is your account balance? -100
Your account is in the red
```

5.2 The if-else statement（if-else 语句）

With the simple if statement you have a choice between executing a statement and skipping it. With an **if-else** you have the choice of executing one or other of two statements.

使用简单的if语句，可以选择执行if后面的语句，或者跳过不去执行，而使用if-else语句，则可以在两条语句中选择一个来执行。

Program Example P5B

```
1  /* Program Example P5B
2     Demonstration of the if-else statement. */
3  #include <stdio.h>
4  int main()
5  {
6    float account_balance ;
7
8    printf( "What is your account balance? " ) ;
9    scanf( "%f", &account_balance ) ;
10   if ( account_balance < 0 )
11     printf( "Your account is in the red\n" ) ;
12   else
13     printf( "Your account is in the black\n" ) ;
14   return 0 ;
15 }
```

In this program, if the value of `account_balance` is less than 0, the `printf()` on line 11 is executed; otherwise the `printf()` on line 13 is executed.

在这个程序中，若变量account_balance的值小于0，则执行第11行语句，否则执行第13行语句。

Now suppose that you want to execute a number of statements rather than just one statement after the `if` and `else`. Braces, i.e. `{` and `}`, are used to enclose a group of statements to form a *compound statement*. This is demonstrated in the next program.

如果if和else后面需要执行的不是一条语句，而是一组语句，那么此时应使用花括号即{和}将if 和else后面的一组语句括起来，从而构成一个**复合语句**。

Program Example P5C

```
1  /* Program Example P5C
2     Using braces to form a compound statement. */
3  #include <stdio.h>
4  int main()
5  {
6    float account_balance, interest ;
7    float overdraft_rate = 10.0 ;
8
9    printf( "What is your account balance? " ) ;
10   scanf( "%f", &account_balance ) ;
11
12   if ( account_balance < 0 )
13   {
14     printf( "Your account is in the red\n" ) ;
15     interest = -account_balance * overdraft_rate / 100.0 ;
16     printf( "The interest charged is %6.2f\n", interest ) ;
17   }
18   else
19   {
20     printf( "Your account is in the black\n" ) ;
21     printf( "There is no interest charged\n" ) ;
22   }
23   return 0 ;
24 }
```

This program will ask you for your bank balance and will then calculate an overdraft charge, assuming an overdraft rate of 10 percent. The balance is tested in line 12. If the balance is less than 0, the statements on lines 14 to 16 enclosed in the braces are executed. If the balance is not less than 0, then the statements on lines 20 and 21 are executed.

这个程序首先要求用户输入账户余额，然后计算透支费用，假设透支利率为10%。程序第12行判断账户余额的正负，如果账户余额小于0，则执行位于花括号内的第14~16行语句，否则执行第20行和第21行语句。

Here is a sample run of this program:

```
What is your account balance? -100
Your account is in the red
The overdraft charge is 10.00
```

The general form of the `if-else` statement is:

```
if ( expression )
{
  statement1
  statement2
  statement3
  ...
  statementn
}
else
{
  statement1
  statement2
  statement3
  ...
  statementn
}
```

The statements within the braces { and } are usually indented. Indentation is optional, but is useful in making the program more readable.

if或else后面位于花括号内的语句序列通常要缩进，虽然缩进不是必须的，但是它有助于提高程序的可读性。

5.3 Logical operators（逻辑运算符）

C provides three logical operators for use in `if` statements as shown in Table 5.2.

C语言提供了三种用于if语句的逻辑运算符。

Table 5.2 logical operators

Logical Operator	Meaning
&&	AND
||	OR
!	NOT

The logical operators `&&` (AND) and `||` (OR) are used to combine tests within an `if` statement.

`&&` is used to join two simple conditions together; the resulting compound condition is only true when *both* simple conditions are true.

If `||` is used to join two simple conditions, the result is true if *either* or *both* are true.

The logical NOT operator `!` is used to test if the result of a condition is *not* true.

逻辑运算符&&（与）、||（或）、!（非）用于if语句条件的组合测试。

&&（与）用于将两个简单条件组合在一起，当且仅当&&两边的条件均为真时，复合判断条件才为真。

||（或）同样用于将两个简单条件组合在一起，只要|| 两边的条件有一个为真，复合判断条件就为真。

Examples:

```
int a = 0, b = 0 ;

if ( a == 0 && b == 0 )    /* An example of && */
  printf( "both a AND b are zero" ) ;

if ( a == 0 || b == 0 )    /* An example of || */
  printf( "a OR b is zero" ) ;
```

```
if ( ! ( a == 0 ) )        /* An example of ! */
  printf( "a is not zero" ) ;

if ( a != 0 ) /* An alternative to the above. */
  printf( "a is not zero" ) ;
```

逻辑非运算符!用于测试if语句的条件是否非真。

5.4 Nested `if` statements（嵌套的 if 语句）

When an `if` statement occurs within another `if` statement it is called a `nested if` statement. For example

如果一个if语句中还包含另一个if语句，则称为**嵌套的**if语句。

```
  if ( a == 0 && b == 0 )
    printf( "Both a AND b are zero" ) ;
```

can be rewritten using a nested `if` as follows:

```
  if ( a == 0 )
  if ( b== 0 )
    printf( "Both a AND b are zero" ) ;
```

As a further example, suppose you want to assign a grade to an examination mark based on the following scheme:

进一步举例，假定要按下面的标准对考试分数划分等级。

Grade	*Mark*
A	70-100
B	60-69
C	50-59
D	40-49
E	30-39
F	0-29

Depending on the exam mark, you want to assign one of the six possible grades.

根据考试分数，可按上面6个等级评定成绩。

Program Example P5D

```
1   /* Program example P5D
2      To demonstrate nested if statements. */
3   #include <stdio.h>
4   int main()
5   {
6     float mark;
7     char grade;
8
9     printf( "Please enter an exam mark ") ;
10    scanf( "%f", &mark ) ;
11
12    if ( mark >= 70 )
13      grade = 'A' ;
14    else
15      if ( mark >= 60 )
16        grade = 'B' ;
17      else
```

```
18        if ( mark >= 50 )
19          grade = 'C' ;
20        else
21          if ( mark >= 40 )
22            grade = 'D' ;
23          else
24            if ( mark >= 30 )
25              grade = 'E' ;
26            else
27              grade = 'F' ;
28
29    printf( "A mark of %4.1f is a grade of %c\n", mark, grade ) ;
30    return 0 ;
31 }
```

The following is a sample run of this program:

```
Please enter an exam mark 55
A mark of 55.0 is a grade of C
```

Note that each `else` is paired with the previous `if`. The optional indentation helps the programmer to see which `else` is intended to go with which `if`.

注意，每一个else都是与其前面最近的一个if进行配对。语句的缩进不是必须的，但缩进有助于程序员观察if与else语句的配对关系，即确定哪一个else与哪一个if配对。

Another version of this program using the **else-if** statement follows in the next example.

Program Example P5E

```
1  /* Program Example P5E
2     Demonstration of the else-if statement. */
3  #include <stdio.h>
4  int main()
5  {
6    float mark ;
7    char grade ;
8
9    printf( "Please enter an exam mark " ) ;
10   scanf( "%f", &mark ) ;
11
12   if ( mark >= 70 )
13       grade = 'A' ;
14   else if ( mark >= 60 )
15       grade = 'B' ;
16   else if ( mark >= 50 )
17       grade = 'C' ;
18   else if ( mark >= 40 )
19       grade = 'D';
20   else if ( mark >= 30 )
21       grade = 'E' ;
22   else grade = 'F' ;
23
```

```
24   printf( "A mark of %4.1f is a grade of %c\n", mark, grade ) ;
25   return 0 ;
26 }
```

5.5 The switch statement（switch 语句）

The **switch** statement can be used as an alternative to a series of `if-else` statements. The next program emulates a four-function calculator. For each calculation you input two numbers and an operator. For example, if you input 5 + 3 then the program will display 8.

有时可用switch语句代替一系列的if-else语句。下面是一个计算器模拟程序，具有加、减、乘、除4种功能，对于每种运算，只要输入两个操作数和一个运算符即可。

Program Example P5F

```
1  /* Program Example P5F
2      A simple four-function calculator.
3      This program demonstrates the use of the switch statement.*/
4  #include <stdio.h>
5  int main()
6  {
7    char operator ;
8    float num1, num2, answer ;
9
10   printf( "Please enter an arithmetic expression (e.g. 1 + 2) " ) ;
11   scanf( "%f%c%f", &num1, &operator, &num2 ) ;
12
13   switch ( operator )
14   {
15     case '+' :
16       answer = num1 + num2 ;
17       printf( "%f plus %f equals %f\n", num1, num2, answer ) ;
18       break ;
19     case '-' :
20       answer = num1 - num2 ;
21       printf( "%f minus %f equals %f\n", num1, num2, answer ) ;
22       break ;
23     case '*' :
24       answer = num1 * num2 ;
25       printf( "%f multiplied by %f equals %f\n",num1,num2,answer ) ;
26       break ;
27     case '/' :
28       answer = num1 / num2 ;
29       printf( "%f divided by %f equals %f\n",num1,num2,answer ) ;
30       break ;
31     default :
32       printf( "Invalid operator\n" ) ;
33   }
34   return 0 ;
35 }
```

A sample run of this program will produce the following:

```
Please enter an arithmetic expression (e.g. 1+2) 5+3
5.000000 plus 3.000000 equals 8.000000
```

Line 11 reads in the two operands and the operator. Note that %c is preceded by a space in the scanf() to read in the value of the operator. (The reason for this is explained after the switch statement is explained below.)

程序的第11行读入两个操作数和一个运算符。注意，这里用scanf( )函数读取运算符时，在%c的前面加了一个空格，原因将在后面讲解switch语句时讲述。

The switch statement is equivalent to a series of if-else statements. The variable or expression to be tested is placed in parentheses after the keyword switch. Unfortunately, this variable or expression can only be of type char or int, which somewhat limits the usefulness of switch.

switch语句相当于一系列的if-else语句，被测试的变量或表达式写在关键字switch后面的括号中，变量或表达式只能是char类型或int类型，这在某种程度上限制了switch语句的应用。

As many cases as are required are enclosed within the braces. Each case begins with the keyword **case**, followed by the value of the variable or expression and ends with a colon. The value of the variable or expression is compared with each case value in turn. If a match is found, then the statements following the matching case are executed.

在switch后面的花括号中可以包含很多分支，每个分支都以关键字case开头，其后跟随一个变量或表达式的值和冒号，switch后括号内的变量或表达式的值依次与每一个case常量值相比较，当发现匹配的case常量时，则执行相应case后面的语句。

Once a match is found and the appropriate statements are executed, the break statement terminates the switch statement. Without the break statement, execution would continue to the end of the switch statement. You can leave out break when you want the same statements executed for several different cases. For example, suppose in the last program you want to use either *, x or X to indicate multiplication. The switch statement would have to be modified as follows:

当发现匹配的case并执行完相应的语句后，使用break语句跳出switch语句，如果没有break 语句，那么程序将依次执行后面的语句，直到switch的最后一条语句为止。当需要对不同的case执行相同的语句时，可以不使用break语句。例如，假设在上面的程序中，可使用字符*、x 或X 来表示乘号，则switch语句可以修改如下。

```
case '*' :
case 'x' :
case 'X' :
  answer = num1 * num2 ;
  printf( "%f multiplied by %f equals %f\n", num1, num2, answer ) ;
  break;
```

Either *, x or X will execute the same statements.

此时，无论用户输入字符*、x或X，都将执行相同的语句。

If no case matches the value of the switch variable, the **default** case is executed. In this program the default case is used to handle an invalid operator.

如果没有找到与switch变量相匹配的case，那么程序将执行default后面的语句。在这个程序中，default用于处理非法运算符。

Now back to why %c is preceded by a space in the scanf() in line 11. The space preceding %c is used so that any whitespace characters the user types in the input expression are ignored. For example, if the user typed 1 +2, the spaces are ignored and the operator is read as +.

现在我们来解释为什么第11 行的scanf( )函数中在%c前面加一个空格。在%c前面加一个空格可以忽略用户输入的表达式中的空白字符。例如，如果用户输入表达式1 +2，那么程序将忽略运算符前面的空格，使得读入的运算符是+。

5.6 The conditional operator ?:（条件运算符）

The conditional operator `?:` is a short form of `if-else`. The following program reads two values from the keyboard and finds the larger of the two using the conditional operator.

条件运算符?:其实就是一种简写形式的if-else语句。下面的程序从键盘输入两个数，然后使用条件运算符求取其中最大的一个。

Program Example P5G

```
1  /* Program Example P5G
2     Demonstration of the conditional operator ?: */
3  #include <stdio.h>
4  int main()
5  {
6   float max, num1, num2 ;
7
8   printf( "Type in two numbers. Press Enter after each number.\n" );
9   scanf( "%f%f", &num1, &num2 ) ;
10
11  /* Assign max to the larger of the two numbers. */
12  max = ( num1 > num2 ) ? num1 : num2 ;
13  printf( "The larger number is %f\n", max ) ;
14  return 0 ;
15 }
```

Line 12 is just a shorthand way of writing

第12行语句其实就是下面语句的简写形式。

```
if ( num1 > num2 )
  max = num1 ;
else
  max = num2 ;
```

A sample run of this program follows:

```
Type in two numbers. Press Enter after each number.
1
2
The larger number is 2.000000
```

There is no easy way to ensure that the user types in numeric data for `num1` and `num2`. Letters rather than numbers could be typed in and the program will carry on regardless. However, some help is at hand, in that `scanf()` returns an integer value to indicate the number of data items successfully read. This value can be used to make a simple error check, as in the following:

没有简单的方法用来确保用户输入给变量num1和num2的值是数值型数据，用户键入的很可能是字符而非数值，而程序却不管这些，仍然继续往下执行。不过，也有一些处理方法，由于scanf( )函数返回的整型值表示其成功读入的数据个数，因此，可以用这个整型值来进行简单的错误检查。

```
numbers_read = scanf( "%f%f", &num1, &num2 ) ;
if ( numbers_read != 2 )
  printf( "Error in reading values from the keyboard" ) ;
else
{
  /* Two numbers have been read from the keyboard. */
  ...
}
```

Programming pitfalls

1. There is no ; immediately after an if statement. For example:

```
if ( account_balance < 0 ) ; /* Misplaced semicolon. */
  printf( "Your account is in the red\n" ) ;
```

should be:

```
if ( account_balance < 0 )
  printf( "Your account is in the red\n" ) ;
```

In the first case the line Your account is in the red is always displayed, regardless of the value in account_balance. The reason for this is that the printf() statement is not controlled by the if. The if only controls the empty statement mistakenly made by the misplaced semicolon.

2. There is no ; after switch.
3. When testing for equality use ==, not =.
4. Each else is matched with the previous if.
5. For each opening brace { there will be a closing brace }.
6. Braces are necessary to control the execution of a block of statements with an if statement.

For example:

```
if ( a == b )
  a = 1 ;
  b = 2 ;
```

In this example, the statement a = 1 is executed only if a and b are equal. However, the statement b = 2 is always executed, regardless of the values of a and b. To execute both statements when a and b are equal you need to use braces:

```
if ( a == b )
{
  a = 1 ;
  b = 2 ;
}
```

7. The logical operators (&& and ||) evaluate the smallest number of operands needed to determine the result of an expression. This means that some operands in the expression may not be evaluated. For example, in

```
if ( a > 1 && b++ > 2 )
```

b++ is evaluated only if the condition a > 1 is logically true.

1. if语句的判断条件的括号后面没有分号。

第一个if语句中错误地在if语句后加了分号，导致无论account_balance的值是什么，屏幕上都显示“Your account is in the red”。原因是if 语句后面的分号使if所控制的分支语句不再是printf( )，而错误地变成空语句了。

2. switch后面没有分号。
3. 测试相等时，应使用逻辑运算符==，而非赋值运算符=。
4. 每个else都与其前面最近的一个if配对。
5. 对于每个左花括号{，都必须有一个右花括号}与之配对。
6. 如果if分支中需要执行的是一个语句块，那么必须使用花括号将其括起来。

在这个例子中，语句“a = 1 ;”仅在a与b相等时才执行，而无论a与b的值是什么，语句“b = 2 ;”都被执行。为了让这两条语句都是在a与b相等时才被执行，应使用花括号将其括起来。

7. 在计算含有逻辑运算符（&&和||）的表达式的值时，仅使用最少数量的操作数来确定该表达式的值。这意味着表达式中的某些操作数可以不必求值。例如在下面的语句中，仅当a > 1为真时，b++才会计算。

Quick syntax reference

	Syntax	Examples

if-else	if (condition)
	{
	  statement(s)
	}
	else
	{
	  statement(s)
	}
		if (n > 0)
		{
		  average = total / n ;
		  printf("%5.2f",average) ;
		}
		else
		  average = 0 ;

?:	variable=(condition)?v1:v2 ;	max = (n1>n2) ? n1:n2 ;

switch	switch (expression)
	{
	case $value_1$:
	    statement(s)
	    break ;
	case $value_2$:
	    statement(s)
	    break ;
	default :
	    statement(s)
	}
		char traffic_light ;
		...
		switch(traffic_light)
		{
		  case 'R':
		  case 'r':
		    printf("Red: STOP") ;
		    break ;
		  case 'G':
		  case 'g':
		    printf("Green: GO") ;
		    break ;
		  case 'A':
		  case 'a':
		   printf("Amber: READY") ;
		   break ;
		  default:
		   printf("FAULT") ;
		}

Exercises

1. Rewrite the following `if-else` using a `switch` statement:

```
if ( marriage_status == 'S' )
   printf( "single" ) ;
else if ( marriage_status == 'M' )
   printf( "married" ) ;
else if ( marriage_status == 'W' )
   printf( "widowed" ) ;
else if ( marriage_status == 'E' )
   printf( "separated" ) ;
else if ( marriage_status == 'D' )
  printf( "divorced" ) ;
else
  printf( "error:invalid code" ) ;
```

2. The following program segment displays an appropriate message depending on the values of three integers n1, n2, and n3.

```
if ( n1== n2 )
{
  if ( n1 == n3 )
  {
    printf ( "n1, n2 and n3 have the same value\n" ) ;
  }
  else
  {
    printf ( "n1 and n2 have the same value\n" ) ;
  }
}
else if ( n1 == n3 )
{
  printf ( "n1 and n3 have the same value\n" ) ;
}
else if ( n2 == n3 )
{
  printf ( "n2 and n3 have the same value" ) ;
}
else
{
  printf( "n1, n2 and n3 have different values");
}
```

   To test the various branches in this code you will need to construct five sets of test data, each set testing one of the branches. Construct the five sets of test data for n1, n2, and n3.
3. Write a program to input two integers from the keyboard and check if the first integer is evenly divisible by the second. (Hint: use the modulus operator %.)
4. Input two numbers from the keyboard and find the smaller of the two using the conditional operator ?:.
5. In a triangle, the sum of any two sides must be greater than the third side. Write a program to input three numbers from the keyboard and determine if they form a valid triangle.
6. Write a program to input a single numeral from the keyboard and displays its value as a word. For example, an input of 5 will display the word "five".
7. Write a program to input a number 1 to 7 from the keyboard, where 1 represents Sunday, 2 Monday, 3 Tuesday, etc. Display the day of the week corresponding to the number typed by the user. If the user types a number outside the range 1 to 7, display an error message.
8. Add the increment operator (I or i) and the decrement operator (D or d) to the simple calculator program P5F.
9. Write a program to input from the keyboard the time of day in Ireland and display the equivalent time in Washington (– 5 hours), Moscow (+ 3 hours), and Beijing (+ 7 hours). Input the time in the 24-hour format, e.g. 22:35 (10:35 p.m.).

10. Write a program to display the effects of an earthquake based on a Richter scale value input from the keyboard. The effects corresponding to a Richter scale value is as follows:

Richter scale value	**Effects**
Less than 4	Little.
4.0 to 4.9	Windows shake.
5.0 to 5.9	Walls crack; poorly built buildings are damaged.
6.0 to 6.9	Chimneys tumble; ordinary buildings are damaged.
7.0 to 7.9	Underground pipes break; well-built buildings are damaged.
More than 7.9	Ground rises and falls in waves; most buildings are destroyed.

Chapter Six
Iterative Control Statements: while, do-while, and for
第 6 章　循环控制语句：while、do-while 和 for

Iterative control statements allow you to execute one or more program statements repeatedly. The C language has three iterative control statements: the **while**, the **do-while** and the **for** statements.

循环控制语句，使我们可以重复地执行程序中的一条或多条语句。C语言提供了三种循环控制语句：while语句、do-while语句和for语句。

6.1 The while statement（while 语句）

The **while** statement causes one or more statements to repeat as long as a specified expression remains true. The general form of the `while` statement is:

while 语句使得只要指定的表达式值为真，就重复地执行一条或多条语句。

```
while ( control expression )
{
  statement_1
  statement_2
  ...
  statement_n
}
```

The statements enclosed within the braces { and } are executed repeatedly while the control expression is true. The repeated execution of one or more program statements is called a *program loop*. The braces enclosing the loop may be omitted if there is only one statement in the loop.

当循环控制表达式为真时，将重复执行花括号内的语句。重复执行花括号内的一条或多条语句，称为**程序循环**。当循环内只有一条语句时，用于将循环括起来的花括号可以省略。

The next program reads in a series of numbers and displays a running total of the numbers. The program stops when the number read is 0.

下面程序的功能是，读入一组数据，然后输出它们的和，当读入的数据为0时，程序结束。

Program Example P6A

```
1  /* Program Example P6A
2     Demonstration of the while statement.
3     This program reads in a series of numbers from the
4     keyboard, prints a running total, and stops when a 0
5     is entered.                                         */
6  #include <stdio.h>
7  int main()
8  {
9    float num, total ;
10
11   total = 0 ;
12   num = 1 ;
13   while ( num != 0 )
```

```
14    {
15      printf( "Please enter a number (0 to finish) " ) ;
16      scanf( "%f", &num ) ;
17      total += num ;
18      printf( "The running total is %f\n\n", total ) ;
19    }
20    printf( "The final total is %f\n", total) ;
21    return 0 ;
22 }
```

A sample run of this program displays the following:

```
Please enter a number 12
The running total is 12.000000

Please enter a number 6.4
The running total is 18.400000

Please enter a number -1.25
The running total is 17.150000

Please enter a number 0
The running total is 17.150000

The final total is 17.150000
```

The statements on lines 15 to 18 are executed repeatedly while the value of the variable `num` is not `0`. When `num` becomes `0`, the loop stops and the statement on line 20 is executed.

当变量num的值不为0 时，将重复执行第15行~第18行的语句。当变量num的值为0时，循环结束，然后执行第20行的语句。

The control expression in a `while` loop is tested *before* the statements in the loop are executed. The sequence in a `while` loop is as follows:

while 循环是在执行循环体内的语句之前先判断循环控制表达式的值。while循环的语句执行顺序如下：

1. Evaluate the control expression.
2. If the control expression is true, execute the statements in the loop and go back to 1.
3. If the control expression is false, exit the loop and execute the next statement after the loop.

1. 计算循环控制表达式的值。
2. 如果循环控制表达式的值为真，则执行循环体内的语句，并返回步骤1。
3. 如果循环控制表达式的值为假，则退出循环，执行循环后面的下一条语句。

It is important to note that if the first evaluation of the control expression is false, the statements in the loop are never executed. This is the purpose of giving the variable `num` a non-zero value in line 12. Line 12 places a value of `1` into `num`, but any non-zero value would do.

尤其值得注意的是，如果第一次计算得到的循环控制表达式的值为假，那么循环体内的语句将一次都不会被执行。这就是在程序第12行为变量num赋一个非0值的原因。第12行语句将变量num赋值为1，但其实为其赋任何非0值都可以。

6.2 The `do-while` loop（do–while 循环）

In a `while` loop the control expression is tested *before* the statements in the loop are executed. The test in a **do-while** loop is done *after* the statements in the loop are executed. This means that the statements in a `do-while` loop are executed at

while循环是在执行循环体内的语句之前先判断循环控制表达式的值。而do-while循环则是先执行循环体内的语句，然后再判断。这就意味着do-while

least once. The sequence in this type of loop is as follows:

1. Execute the statements in the loop.
2. Evaluate the control expression.
3. If the control expression is true then go back to 1.
4. If the control expression is false then exit the loop and execute the next statement after the loop.

循环中的语句将至少执行一次。do-while循环的执行顺序是

1. 执行循环体内的语句。
2. 计算循环控制表达式的值。
3. 如果循环控制表达式的值为真，则转至步骤1。
4. 如果循环控制表达式的值为假，则退出循环，执行循环后面的下一条语句。

The general form of the do-while loop is:

```
do
{
  statement1
  statement2
  ...
  statementn
}
while ( control expression ) ;
```

The next program replaces the while loop in program P6A with a do-while loop.

Program Example P6B

```
1   /* Program Example P6B
2      Demonstration of the do-while loop.
3      This program reads a series of numbers from the
4      keyboard, prints a running total, and stops when a 0
5      is entered.                                              */
6   #include <stdio.h>
7   int main()
8   {
9     float num, total ;
10
11    total = 0 ;
12    do
13    {
14      printf( "Please enter a number (0 to finish) " ) ;
15      scanf( "%f", &num ) ;
16      total += num ;
17      printf( "The running total is %f\n\n", total ) ;
18  }
19    while ( num != 0 ) ;
20    printf( "The final total is %f\n", total ) ;
21    return 0 ;
22  }
```

Using a do-while loop means there is no need to initialise the variable num, because the loop is executed at least once. The output from this program and program P6A are the same.

由于do-while循环的循环体至少执行一次，因此这里无须再对变量num进行初始化。

6.3 The `for` statement（for 语句）

The **for** statement is used to execute one or more statements a specified number of times. The general format of the `for` statement is:

for语句通常用于一条或多条语句重复执行指定的次数的情况。

```
for (initial expression; continue condition; increment expression)
{
  statement₁
  statement₂
  ...
  statementₙ
}
```

The `for` statement consists of three expressions enclosed in parentheses and separated by semicolons.

The `initial expression` is executed once at the beginning of the loop.

The loop continues while the `continue expression` is true (a non-zero value) and terminates when the `continue expression` becomes false (a zero value). The test occurs before each pass, including the first pass, through the loop.

The `increment expression` is executed at the end of every pass through the loop.

The braces `{` and `}` are needed only when there is more than one statement in the loop.

在for语句中关键字for后的括号中包含3个表达式，它们之间是用分号分隔的。

第1个表达式initial expression是对循环控制变量进行初始化，它只在循环开始时执行一次。

第2表达式continue expression是循环继续的条件，若其为真（非0值），则循环继续，否则当其为假（0值）时，将终止循环。在包括第一次执行循环的整个循环中，每次执行循环时都要进行这种测试。

第3个表达式increment expression是对循环控制变量进行增值，每执行完一次循环体都要对循环控制变量进行一次增值。

当循环体中的语句多于一条语句时，要使用花括号将它们括起来。

Program Example P6C

```
1  /* Program Example P6C
2     Demonstration of a loop using the for statement.
3     This program displays the numbers 1 to 10.      */
4  #include <stdio.h>
5  int main()
6  {
7    int i ;
8
9    for ( i= 1 ; i <= 10 ; i++ )
10   {
11     printf( "%d\n", i ) ;
12   }
13   return 0 ;
14 }
```

The body of the loop is contained within the braces on lines 10 and 12. The loop contains only one statement, so the braces may be omitted. However, the braces are useful in that they clearly show the body of the loop.

The `for` statement on line 9 causes the loop to be executed ten times, with the variable `i` starting at `1` and continuing while the value of `i` is less than or equal to `10`. Each time the loop is completed the value of `i` is incremented by 1. When the value of i becomes 11 the loop terminates.

程序的第10行~第12行是循环体，由于它只包含一条语句，因此其外侧的花括号可以省略。但是，添加花括号有助于突出循环体。

There are many variations you can add to the simple for on line 9. Try modifying the program in each of the following ways:

1. To display the numbers from 10 down to 1, change line 9 to:

   ```
   for ( i = 10 ; i > 0 ; i -- )
   ```

   Here i starts at 10 and is decremented at the end of each pass through the loop until it eventually becomes 0.
2. To display the even numbers from 2 to10, change line 9 to:

   ```
   for ( i = 2 ; i <= 10 ; i += 2 )
   ```

   In this for statement the variable i is initialised to 2 and increases by 2 each time through the loop, displaying the numbers 2 4 6 8 10.

Either initial expression or increment expression, or both, may consist of multiple expressions separated by commas. For example:

在变量初始化表达式和变量增值表达式中，可以包含多个表达式，使用逗号分隔即可。

```
for ( i = 0, j = 0 ; i < 10 ; i++, j++ )
```

This loop initialises both i and j to 0 and increments both of them at the end of each pass through the loop.

Any or all of the three expressions may be omitted from a for statement, but the two semicolons must always be present in the statement. For example, you can omit the initialisation of i in line 9 of program P6C and perform it in line 7 instead.

for语句中的3个表达式都可以省略，但是括号中的两个分号不能省略。例如，可以将程序P6C第9行的for语句中对变量i初始化的表达式省略。

```
7 int i = 0 ;

9 for ( ; i <= 10 ; i++ )
```

Note that the statement for(;;) will create an infinite loop, because there is no condition to end the loop.

The next program asks the user for a number and displays a table of squares and cubes from 1 up to the number specified.

注意语句for(; ;)将导致无限循环，这是因为其中没有给出使循环终止的条件。

下面程序的功能是，要求用户输入一个数字，然后在一个表格中输出从1到该值之间的所有数字的平方值和立方值。

Program Example P6D

```
1  /* Program Example P6D
2     This program displays a table of squares and cubes. */
3  #include <stdio.h>
4  int main()
5  {
6    int num, i ;
7
8    printf( "Please enter a number " ) ;
9    scanf( "%d", &num ) ;
10
11   printf( "Number   Square   Cube\n" ) ;
12   printf( "-----------------------\n" ) ;
13   for ( i = 1 ; i <= num ; i++ )
14   {
```

```
15     printf( "%4d    %4d    %4d\n", i, i*i, i*i*i ) ;
16   }
17   return 0 ;
18 }
```

A sample run of this program is:

```
Please enter a number 5

Number   Square   Cube
----------------------
   1        1        1
   2        4        8
   3        9       27
   4       16       64
   5       25      125
```

6.4 Nested loops（嵌套的循环）

A `for` loop can contain any valid C statements, including another `for` loop. When a loop is contained within another loop it is called a *nested loop*. The next program displays a 12 by 12 multiplication table using a nested loop.

在for循环体中可以包含任何合法的C语句，包括另一个for循环。当一个循环体中包含另一个循环时，则称这个循环为**嵌套的循环**。

下面的程序用一个嵌套的循环显示一个12 × 12的乘法表。

Program Example P6E

```
1  /* Program Example P6E
2     This program displays a 12 by 12 multiplication table
3     using a nested loop.                                       */
4  #include <stdio.h>
5  int main()
6  {
7    int i, j ;
8
9    printf( "    " ) ;
10   for ( i = 1 ; i <= 12 ; i++ )
11   {
12     printf( "%5d", i ) ;
13   }
14   printf( "\n   +" ) ;
15   for ( i = 0 ; i <= 60 ; i++ )
16   {
17     printf( "-" ) ;
18   }
19   for ( i = 1 ; i <= 12 ; i++ )  /* Start of outer loop   <--+ */
20   {                              /*                          | */
21     printf( "\n%2d |", i ) ;     /*                          | */
22     for( j = 1 ; j <= 12 ; j++ ) /* Start of inner loop <--+  | */
23     {                            /*                      |  | */
24       printf( "%5d", i*j ) ;     /*                      |  | */
25     }                            /* End of inner loop  <--+  | */
26   }                              /* End of outer loop     <--+ */
```

```
27    printf( "\n" ) ;
28   return 0 ;
29 }
```

This program will display the following table:

```
         1    2    3    4    5    6    7    8    9   10   11   12
  +-------------------------------------------------------------
1 |      1    2    3    4    5    6    7    8    9   10   11   12
2 |      2    4    6    8   10   12   14   16   18   20   22   24
3 |      3    6    9   12   15   18   21   24   27   30   33   36
4 |      4    8   12   16   20   24   28   32   36   40   44   48
5 |      5   10   15   20   25   30   35   40   45   50   55   60
6 |      6   12   18   24   30   36   42   48   54   60   66   72
7 |      7   14   21   28   35   42   49   56   63   70   77   84
8 |      8   16   24   32   40   48   56   64   72   80   88   96
9 |      9   18   27   36   45   54   63   72   81   90   99  108
10|     10   20   30   40   50   60   70   80   90  100  110  120
11|     11   22   33   44   55   66   77   88   99  110  121  132
12|     12   24   36   48   60   72   84   96  108  120  132  144
```

The loop in lines 10 to 13 displays the numbers 1 to 12 across the screen, and the loop in lines 15 to 18 displays the hyphens beneath them. The remainder of the program uses a nested loop to display the numbers in the table.

The outer loop (lines 19 to 26) starts with `i` at `1`. Line 21 displays a `1` and the vertical stroke character | at the left of the screen.

The inner loop (lines 22 to 25) is then executed to completion, with `j` starting at `1` and ending when `j` exceeds `12`. Each iteration of the inner loop displays a number in the multiplication table.

When the inner loop is completed, the outer loop regains control and `i` is incremented to 2.

Line 21 then displays `2 |` at the left of the screen, and the inner loop on lines 22 to 25 is executed again.

The program continues until the outer loop is completed when the value of `i` exceeds `12` and the program terminates.

Programming pitfalls

1. When you open a block of statements with an opening brace `{`, you must close the block with a closing brace `}`.

2. There is no semicolon immediately after the `while` or `for` statements. For example:

```
int i ;

for ( i = 0 ; i < 10 ; i++ ) ; /* Misplaced semicolon. */
  printf( "The value of i is %d\n", i ) ;
```

This loop does not contain any statements and will not display the values of `i` from 0 to 9, as expected. Only the final value of `i` (i.e. 10) will be displayed.

3. Be careful in specifying the terminating condition in a `for` loop. For example:

```
int i ;

for ( i = 0 ; i == 10 ; i++ ) /* This loop does nothing. */
  printf( "The value of i is %d\n", i ) ;
```

This loop does nothing, because `i == 10` is false at the start of the loop (`i` is in fact `0`) and the loop terminates immediately. Replace `i == 10` with `i < 10` and the loop will execute ten times.

4. There is no semicolon after `while` in a while loop, but there is in a `do-while` loop. See line 13 of program P6A and line 19 of program P6B.

5. There is a limit to the precision with which floating-point numbers are represented. This is important when testing a floating-point number for equality in an `if` or in a `for` loop. For example, consider the following loop:

```
float f ;

for ( f = 0.0 ; f != 1.1 ; f += 0.1 )
{
  /* Statement(s) in the loop. */
}
```

On most computers this will result in an infinite loop. The reason for this is that `f` may never equal `1.1` exactly. You can allow for this situation by writing the loop as:

```
for ( f = 0.0 ; f <= 1.1 ; f += 0.1 )
{
  /* Statement(s) in the loop. */
}
```

1. 在对语句块使用花括号时，注意每一个左花括号{ 都必须与一个右花括号} 成对出现。
2. 紧随while语句或for语句的后面没有分号。
3. 在for循环中，要小心指定循环终止条件。
4. 在while循环中，while的后面没有分号，但在do-while循环中，while的后面有分号。
5. 浮点数存在精度限制，这对于在if语句或for语句中检测浮点数相等是非常重要的。

Quick syntax reference

	Syntax	Examples

while	while (condition)
	{
	  statement(s)
	}
		/* Read and total until n is 0. */
		int n = 1 ;
		int total = 0 ;
		while (n != 0)
		{
		  scanf("%d", &n) ;
		  total += n ;
		}

do-while	do
	{
	  statement(s)
	}
	while (condition) ;
		/* Read and total until n is 0. */
		int n ;
		int total = 0 ;
		do
		{
		  scanf("%d", &n) ;
		  total += n ;
		}
		while (n != 0) ;

for	for(initial expression;
	    continue condition;
	    increment expression)
	{
	  statement(s)
	}
		/* Read and total 10 numbers. */
		int n ;
		int total = 0 ;
		for (i=0 ; i<10 ; i++)
		{
		  scanf("%d", &n) ;
		  total += n ;
		}

Exercises

1. What is the output from the following program?

```
int main()
{
  int i, j = 1 ;
  for ( i = 10 ; i > 0 ; i /= 2, j++ )
    printf( "%d %d\n", i, j ) ;
  return 0 ;
}
```

2. Modify program P6A to calculate the average along with the total of the numbers entered.
3. Rewrite the following using a `for` loop.

```
int total = 0 ;
int i = 0 ;

while ( i < 10 )
{
  scanf ( "%d", &n ) ;
  total += n ;
  i++ ;
}
```

4. What is displayed when the following program is run and the number 1234 is entered?

```
#include <stdio.h>
int main()
{
  int num ;
  printf( "Please enter a number " ) ;
  scanf( "%d", &num ) ;
  do
  {
    printf( "%d", num % 10 ) ;
    num /= 10 ;
  }
  while ( num != 0 ) ;
  return 0 ;
}
```

5. Write a program that allows a teacher to enter a percentage mark for each student in a class. The teacher enters a negative mark to indicate that there are no more marks to be entered. Once all the marks have been entered, the program displays the average percentage mark for the class.
6. Write a program to find the sum of all the odd integers in the range 1 to 99.
7. Write a program to display all the hour and minute values in a 24-hour clock, i.e.
   0:01 0:02 ⋯ 12:59 0:00.
   How would you display the values in fifteen-minute intervals?
8. Write a program to display a Christmas tree.

The tree is consists of a series of tiers of increasing size. There are three tiers in the tree above. The program inputs the number of tiers from the keyboard.

Chapter Seven
Arrays
第 7 章 数 组

7.1 Introduction to arrays（引言）

Consider the problem of writing a C program to input the ages of ten people and to calculate their average age. You could use `scanf()` to input the age of each person into ten different variables, total the variables, and calculate the average. The next program calculates the average using this method.

考虑一下，如何编写一个C程序来读取10个人的年龄，然后计算它们的平均值？当然，我们可以用scanf( )函数依次输入10个人的年龄，将其保存到10个变量中，对它们进行求和，并计算其平均值。下面的程序就是使用这种方法来计算平均值的。

Program Example P7A

```
1  /* Program Example P7A
2     This program calculates the average age of ten people
3     using ten different variables.                      */
4  #include <stdio.h>
5  int main()
6  {
7    int age1, age2, age3, age4, age5,
8        age6, age7, age8, age9, age10 ;
9    int total_age ;
10
11   printf( "Please enter the ages of ten people\n" ) ;
12   /* Input each age from the keyboard. */
13   scanf( "%d", &age1 ) ;
14   scanf( "%d", &age2 ) ;
15   scanf( "%d", &age3 ) ;
16   scanf( "%d", &age4 ) ;
17   scanf( "%d", &age5 ) ;
18   scanf( "%d", &age6 ) ;
19   scanf( "%d", &age7 ) ;
20   scanf( "%d", &age8 ) ;
21   scanf( "%d", &age9 ) ;
22   scanf( "%d", &age10) ;
23
24   total_age = age1 + age2 + age3 + age4 + age5 +
25               age6 + age7 + age8 + age9 + age10 ;
26
27   printf( "The average of %d %d %d %d %d %d %d %d %d %d is %d\n",
28           age1,age2,age3,age4,age5,age6,age7,age8,age9,age10,
29           total_age / 10) ;
30
31   return 0 ;
32 }
```

Here is a sample run of this program:

```
Please enter the ages of ten people
41
67
21
7
59
57
41
74
47
68
The average of 41 67 21 7 59 57 41 74 47 68 is 48
```

The ages are input on lines 13 to 22. Lines 24 to 25 calculate the total ages. Lines 27 to 29 display the ages and calculates and displays the average (without the decimal portion).

Writing this program is manageable for ten people, but what happens if the number of people increases to a hundred or more? Do you need another ninety variables? This is where *arrays* are useful.

An *array* is a group of variables of the same data type, such as ten `int`s, fifteen `char`s or a hundred `float`s. With an array, the ten ages are held together in memory under one name, i.e. the array name.

Each individual element of the array is accessed by reference to its position in the array relative to the first element of the array. The position of an element in an array is called the *index* or *subscript*. The first element in an array of ten elements has an index value of 0, and the last element has an index value of 9. Note that the index goes from 0 to 9, not 1 to 10. To refer to a particular element of an array you give the array name and the index in brackets. In the array `ages`, shown in Figure 7.1 above, the first element is `ages[0]` and has the value 41, the second element is `ages[1]` and has the value 67, and so on. The tenth element is `ages[9]` and has the value 68.

数组是一组具有相同数据类型（例如10个整型数据，15个字符型数据，或是100个浮点型数据）的变量的集合。每个数组元素是通过引用其相对于数组第一个元素的位置来存取的。数组元素在数组中的位置，称为数组的**索引**或**下标**。对于有10个元素的数组，它的第一个元素的下标值是0，最后一个元素的下标值是9。注意，数组的下标值是从0~9，而不是从1到10。可以通过给定数组名及写在方括号中的下标值来引用数组中的某个特定的元素。

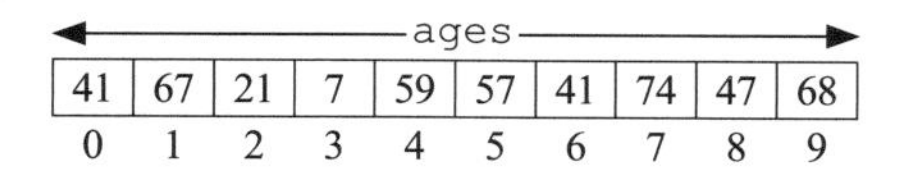

Figure 7.1 ages array

The previous program, P7A, can be rewritten using an array:

Program Example P7B

```
1  /* Program Example P7B
2     This program calculates the average age of ten people
3     using an array.                                     */
4  #include <stdio.h>
5  int main()
6  {
```

```
7     int ages[10] ;
8     int i ;
9     int total_age = 0 ;
10
11    printf( "Please enter the ages of ten people\n" ) ;
12    /* Input and total each age. */
13    for ( i = 0 ; i < 10 ; i++ )
14    {
15      scanf( "%d", &ages[i] ) ;
16      total_age += ages[i] ;
17    }
18
19    printf( "The average of ");
20    for ( i = 0 ; i < 10 ; i++ )
21      printf( "%d ", ages[i] ) ;
22    printf( "is %d\n", total_age / 10 ) ;
23    return 0 ;
24 }
```

Here is a sample run of this program:

```
Please enter the ages of ten people
41
67
21
7
59
57
41
74
47
68
The average of 41 67 21 7 59 57 41 74 47 68 is 48
```

The statement

```
int ages[10] ;
```

defines `ages` as an array of ten integers. An array is defined by stating the type of its elements, its name, and the number of elements in the array. In general, the format is:

数组是通过声明数组元素的类型、数组名和数组元素的个数来定义的。

```
type variable_name[number_of_elements] ;
```

For example:

```
float array_1[50] ; /* An array of 50 floats. */
double array_2[20] ; /* An array of 20 doubles. */
```

In program P7B, if the number of people is increased from ten to a hundred, then you must change all the occurrences of 10 to 100. This is a lot easier than introducing ninety more variables, as you would have to do in program P7A.

The for loop in lines 13 to 17 of program P7B is used to read in a value for each of the elements of the array and add them to total_age.
The variable i is used to hold the value of the index or subscript of the array. On the first pass through the loop, i has the value 0, and line 15 reads in a value for ages[0], i.e. the first element. Line 16 then adds ages[0] to total_age.
On the second pass through the loop, i has a value of 1, line 15 reads in a value for ages[1], and line 16 adds ages[1] to total_age. The loop continues until the tenth element, ages[9], is read in and added to total_age.
The value of each element in the array ages is displayed using the for loop on lines 20 to 21 and the average is displayed on line 22.
The next program further demonstrates arrays by reading in ten ages from the keyboard and displaying the youngest, the oldest and the average age.

变量i的作用是保存数组的下标值或索引值，在第一轮for循环中，变量i被初始化为0，因此程序第15行语句读入的数据保存到数组元素ages[0]即数组的第一个元素中，第16行语句将ages[0]的值累加到变量total_age 中。
在第二轮for循环中，变量i的值变为1，第15行语句读入的数据保存到数组元素ages[1]中，第16行语句将ages[1]的值累加到变量total_age中。重复上述循环，直到第10个数组元素ages[9]的值被读入并累加到变量total_age中时，结束循环。

Program Example P7C

```
1  /* Program Example P7C
2     This program inputs a series of ages from the keyboard
3     and displays the youngest, the oldest and the average.  */
4  #include <stdio.h>
5  #define SIZE 10
6  int main()
7  {
8    int ages[SIZE] ;
9    int i ;
10   int total_age = 0 ;
11   int youngest, oldest ;
12
13   printf( "Please enter the ages of %d people\n", SIZE ) ;
14   /* Input and total each age. */
15   for ( i = 0 ; i < SIZE ; i ++ )
16   {
17     scanf( "%d", &ages[i] ) ;
18     total_age += ages[i] ;
19   }
20
21   youngest = ages[0] ;
22   oldest = ages[0] ;
23
24   for ( i = 0 ; i < SIZE ; i ++ )
25   {
26     if ( ages[i] > oldest )
27     {
28       oldest = ages[i] ;
29     }
30     if ( ages[i] < youngest )
```

```
31      {
32        youngest = ages[i] ;
33      }
34    }
35
36    printf( "The youngest is %d\n", youngest) ;
37    printf( "The oldest is %d\n", oldest ) ;
38    printf( "The average is %d\n", total_age / SIZE ) ;
39    return 0 ;
40 }
```

A sample run of this program is:

```
Please enter the ages of ten people
41
67
21
7
59
57
41
74
47
68
The youngest is 7
The oldest is 74
The average is 48
```

This program uses the preprocessor directive `#define`. The `#define` directive is used to assign a *symbolic name* to a constant. The symbolic name is then used throughout the program in place of the constant itself. In line 5 of this program, the symbolic name `SIZE` is assigned to the constant `10`. Using a symbolic name makes the program easier to maintain. For example, if you wanted the above program to allow for twenty ages rather than ten, you only have to change line 5 to:

这个程序中使用了编译预处理指令#define，它的作用是为常量定义一个**符号名**，在程序中可以使用该符号名替代常量本身。在程序的第5行，程序为常量10定义了一个符号名SIZE，使用符号名代替常量可以使程序更易于维护。按照惯例，为了与变量名相区分，常量的符号名通常采用大写字母来表示。

```
#define SIZE 20
```

By convention, symbolic names are written in uppercase to distinguish them from variable names.

The `for` loop in lines 15 to 19 reads in values into the array `ages` and totals them in `total_age`.

Lines 21 and 22 assign the first element of the array to the variables `youngest` and `oldest`. The `for` loop in lines 24 to 34 compares each element in the array with the values in the variables `youngest` and `oldest`. When an element larger than `oldest` is found, the value of this element is assigned to `oldest`. When an element is found that is less then `youngest`, this element is assigned to `youngest`. When the loop is completed, the smallest element of the array is in `youngest` and the largest is in `oldest`.

程序第21行和第22行用数组第一个元素的值为变量youngest和oldest初始化，程序第24行~第34行的for循环的作用是，将数组的每个元素与变量youngest和oldest的值进行比较，当某个数组元素值大于变量oldest的值时，就用其替换变量oldest的值，同样，当某个数组元素值小于变量youngest的值时，就用其替换变量youngest的值。

The variables `oldest` and `youngest` were initially assigned to the first element of the array in lines 21 and 22. In fact you could assign them to any element of the array and the program would still work. Try it!

当循环结束时，变量youngest的值即为数组元素中的最小值，而变量oldest的值即为数组元素中的最大值。

7.2 Initialising arrays（数组初始化）

The next program demonstrates array initialisation by asking the user to enter a month and displaying the number of days in that month (leap years excepted).

下面的程序演示了数组的初始化方法，程序首先请用户输入一个月份值，然后显示这个月份对应的天数（不考虑闰年的情况）。

Program Example P7D

```
1   /* Program Example P7D
2      To display the number of days in a month. */
3   #include <stdio.h>
4   #define NO_OF_MONTHS 12
5   int main()
6   {
7     int days[ NO_OF_MONTHS ] =
8             { 31, 28, 31, 30, 31, 30, 31, 31, 30, 31, 30, 31 } ;
9     int month ;
10
11    printf( "Please enter a month (1 = Jan., 2 = Feb., etc.) " ) ;
12    do
13    {
14      scanf( "%d", &month ) ;
15    }
16    while ( month < 1 || month > 12 ) ;
17
18    printf( "\nThe number of days in month %d is %d\n",
19            month, days[month-1] ) ;
20    return 0 ;
21  }
```

A sample run of this program is:

```
Please enter a month (1 = Jan., 2 = Feb., etc.) 9
The number of days in month 9 is 30
```

Lines 7 and 8 of this program define and initialise an array called `days`. The initial values are separated by commas and placed between braces. When the list of initial values is less than the number of elements in the array, the remaining elements are initialised to 0. For example,

当初始化列表中的初值个数少于数组元素的个数时，剩余的未赋初值的数组元素将自动初始化为0值。

```
float values[5] = { 2.3, 7.8, 1.3 } ;
```

initialises the first three elements with the values specified within the braces. The remaining two elements of the array are initialised to 0.

If an array is defined without specifying the number of elements and is initialised to a series of values, the number of elements in the array is taken to be the same as the number of initial values. This means that

如果定义数组时没有指定数组元素的个数，但为数组指定了一定数量的初值，那么系统默认为数组元素的个数与给定的初值个数相同，这意味着下面两条定义数组numbers的语句是等价的。

```
int numbers[] = { 0, 1, 2, 3, 4, 5, 6, 7, 8 } ;
```

and

```
int numbers[9] = { 0, 1, 2, 3, 4, 5, 6, 7, 8 } ;
```

are equivalent definitions of the array `numbers`.

7.3 Two-dimensional arrays（二维数组）

So far, we have only used one-dimensional arrays, i.e. arrays with just one row of elements. A two-dimensional array has more than one row of elements. For example, to record the number of students using one of five computer laboratories over a week, the data could be recorded in the following table:

到目前为止，我们使用的都是一维数组，所谓一维数组是指只有一行元素的数组，而二维数组可以拥有超过一行的元素。例如，可以使用下面的表格记录一周内使用5个计算机实验室的学生人数。

Computer laboratory number

Day 1	1	2	3	4	5
Day 2	120	215	145	156	139
Day 3	124	231	143	151	136
Day 4	119	234	139	147	135
Day 5	121	229	140	151	141
Day 5	110	199	138	120	130
Day 6	62	30	37	56	34
Day 7	12	18	11	16	13

This table has a row for each day of the week and a column for each computer laboratory. This is an example of a two-dimensional array. To define two-dimensional arrays, enclose the size of each dimension of the array in square brackets. For example,

这个表格的行代表一周中的某一天，列代表某个计算机实验室，这就是二维数组应用的一个实例。定义一个二维数组时，需指定数组每一维的长度，并且用方括号将它们括起来。

```
int usage[7][5] ;
```

defines an integer array of seven rows and five columns.

To access an element of a two-dimensional array, you specify the row and the column. Note that the row number starts at 0 and ends at 6, and the column number starts at 0 and ends at 4. For example:

为了访问二维数组中的元素，需要指定该元素在数组中所处的行和列的位置。

```
usage[0][0] is 120        i.e. row 0, column 0
usage[0][4] is 139             row 0, column 4
usage[6][0] is  12             row 6, column 0
and
usage[6][4] is  13             row 6, column 4
```

The row number is in the first set of square brackets and the column number is in the second set of square brackets.

数组元素前面方括号中的值表示行号，后面方括号中的值表示列号。

The next program reads in the number of students using the five laboratories over seven days into a two-dimensional array `usage`. The program then calculates the average usage for each laboratory. Run this program and study the code to see how it works.

Program Example P7E

```
1  /* Program Example P7E
2     This program reads in the number of students using 5 computer
3     labs over 7 days and displays the average usage for each lab.*/
4  #include <stdio.h>
5  #define NO_OF_DAYS 7
6  #define NO_OF_LABS 5
7  int main()
8  {
9    int usage[ NO_OF_DAYS ][ NO_OF_LABS ] ;
10   int day, lab, total_usage, average ;
11
12   /* Read each lab's usage for each day. */
13   for ( day = 0 ; day < NO_OF_DAYS ; day++ )
14   {
15     printf( "Enter the usage for day %d\n", day + 1 ) ;
16     for ( lab = 0 ; lab < NO_OF_LABS ; lab++ )
17     {
18       printf( "Lab number %d ", lab + 1 ) ;
19       scanf( "%d", &usage[day][lab] ) ;
20     }
21   }
22
23   /* Calculate the average usage for each laboratory. */
24   for ( lab = 0 ; lab < NO_OF_LABS ; lab++ )
25   {
26     total_usage = 0 ;
27     for ( day = 0 ; day < NO_OF_DAYS ; day++ )
28     {
29       total_usage += usage[day][lab] ;
30     }
31     average = total_usage / NO_OF_DAYS ;
32     printf( "\nLab number %d has an average usage of %d\n",
33             lab+1, average ) ;
34   }
35   return 0 ;
36 }
```

7.4 Initialising two-dimensional arrays（二维数组的初始化）

Two-dimensional arrays are initialised by enclosing the initial values in braces, as for one-dimensional arrays. For example,

```
int vals[4][3] = { 4, 9, 5, 2, 11, 3, 21, 9, 7, 10, 1, 8 } ;
```

initialises the first row of `vals` with `4`, `9`, and `5`. The second row is initialised with `2`, `11`, and `3`. The third row is initialised with `21`, `9`, and `7` and the fourth row is initialised with `10`, `1`, and `8`.

与一维数组的初始化方法类似，二维数组的初始化也是将初值写入花括号中赋给数组元素。

Readability is improved if you place the initial values of each row on a separate line, as follows:

```
int vals[4][3] = { 4, 9, 5,
                   2, 11, 3,
                   21, 9, 7,
                   10, 1, 8 } ;
```

Additional braces may also be used to separate the rows, as follows:

```
int vals[4][3] = { { 4, 9, 5 },
                   { 2, 11, 3 },
                   { 21, 9, 7 },
                   { 10, 1, 8 } } ;
```

As with one-dimensional arrays, you can omit the number of rows and let the compiler calculate the number of rows from the initial values enclosed in the braces. Therefore, you can rewrite the above definition of `vals` as:

和一维数组类似，对二维数组进行初始化时，可以不声明数组的行数，编译器会自动根据花括号中给出的初值个数计算二维数组的行数。因此，可以按如下方法对二维数组进行定义。

```
int vals[][3] = { { 4, 9, 5 },
                  { 2, 11, 3 },
                  { 21, 9, 7 },
                  { 10, 1, 8 } } ;
```

As with one-dimensional arrays, missing values are initialised to 0. For example, the definition

与一维数组初始化类似，如果初值数量小于数组长度，那么未赋初值的数组元素将初始化为0值。

```
int vals[4][3] = { { 4,  9 },
                   {  7  } } ;
```

will result in

`vals[0][0] = 4, vals[0][1] = 9` and `vals[1][0] = 7`

with all the remaining elements being 0. Note that for an array of 4 rows and 3 columns, the number of rows is required here; otherwise C would assume it to be 2, since there are only two rows of initial values.

7.5 Multi-dimensional arrays（多维数组）

C does not restrict you to two-dimensional arrays. You can define arrays with any number of dimensions. For example, if in program P7E you wanted to store the usage of the five laboratories for each day of the fifty-two weeks of a year, the array `usage` would be defined as:

在C语言中，数组维数并不只限制在二维，我们可以定义任意维数的数组。

```
#define NO_OF_WEEKS 52
#define NO_OF_DAYS 7
#define NO_OF_LABS 5

int usage[NO_OF_WEEKS][NO_OF_DAYS][NO_OF_LABS] ;
```

The elements of this array are accessed by using three subscripts. For example, `usage[0][2][4]` is the usage in the first week, day 3 in laboratory number 5.

Programming pitfalls

1. There is no semicolon after a `#define`. For example:

```
#define SIZE 10 ;        /* Invalid! */
```

2. There is no = in a `#define`. For example:

```
#define SIZE = 10        /* Invalid! */
```

should be

```
#define SIZE 10
```

3. The number of element in each dimension of an array is placed between brackets [ ] and not between parentheses ( ).
4. The range of a subscript is 0 to the number of elements in an array less one. It is a common error to define an array with, for example, ten elements and then attempt to use a subscript value of 10. The subscripts in this case range from 0 to 9. For example:

```
int i, a[10] ;

for ( i = 0 ; i <= 10 ; i++ )
  a[i] = 0 ;
```

This may cause an infinite loop. When `i` is `10`, `a[i]` is assigned `0`. However, `a[10]` does not exist, so `0` is stored in the memory location immediately after `a[9]`. If the variable i happens to be stored after `a[9]`, then i becomes `0`, and so the loop starts again.

1. 在#define后面不应有分号。
2. 在#define 中不应有赋值运算符=。
3. 数组每一维的大小应写在方括号[ ]中，不应写在圆括号( )中。
4. 数组的下标范围是从0 开始到数组元素个数减1。例如，定义一个具有10个元素的一维数组，其正确的下标范围是从0~9，使用下标10去访问数组中第10 个元素是一个常见的错误。这条语句有可能导致死循环。这是因为，当i值为10时，程序将a[10]赋值为0，而a[10]本身并不存在，系统只会将0 值存储到a[9]后紧挨着它的第一个内存单元中，而如果恰巧变量i存储在a[9]后的第一个内存单元中，那么变量i的值将变为0，使其满足循环继续条件，从而导致循环重新开始。

Quick syntax reference

	Syntax	Examples
Defining arrays	`type array[`N_1`][`N_2`]...[`N_n`] ;` N_1,N_2...N_n are the number of elements in each dimension of the array. N_1,N_2...N_n are positive constant integers, not variables.	`int a[10] ;` `float b[5][9] ;`

Array subscripts	`array[`i_1`][`i_2`]...[`i_n`]`
	indexes or subscripts i_1,i_2...i_n are integer constants or variables.
		`a[0] /* 1st  element  */`
		`a[9] /* 10th element  */`
		`b[0][0] /* Row 1,col 1 */`
		`b[4][8] /* Row 5,col 9 */`

Exercises

1. What are the subscript ranges of the following arrays?

(a) `int array1[6] ;`

(b) `float array2[] = { 1.3, 2.9, 11.8, 0 } ;`

(c) `int array3[6][3] ;`

(d) `int array4[][4] = { { 6, 2, 1, 3 } , { 7, 3, 8, 1 } } ;`

2. Write statements to define each of the following:
   (a) a one-dimensional array of floating-point numbers with ten elements
   (b) a one-dimensional array of characters with five elements
   (c) a two-dimensional array of integers with seven rows and eight columns
   (d) a 10 by 5 two-dimensional array of double precision numbers
   (e) a 10 by 8 by 15 three-dimensional array of integers
3. What is the output from the following program segment?

```
int i, c1 = 0, c2 = 0 ;
int a[] = { 6,7,3, 13, 11, 5, 1, 15, 9, 4 } ;
for ( i = 0; i < 10; i++ )
{
  if( i%2 == 0 )
    c1++ ;
  if ( a[i]%2 == 0 )
    c2++ ;
}
printf( "c1=%d c2=%d\n", c1, c2 ) ;
```

4. Write a program to read in fifteen numbers from the keyboard and display them as follows:
   (a) each number on a separate line
   (b) on one line, each number separated by a single space
   (c) as in (b) but in the reverse order to which they were input
5. Write a program to input numbers to two one-dimensional arrays, each having five elements, and display the result of multiplying corresponding elements together.
6. The number of users logging into a network every hour is input from the keyboard into a 24-element integer array. Write a program to display a report of the form:

```
        Time               Number of logins        Percentage of total
    0:00 - 1:00                   1                        0.3
    1:00 - 2:00                   2                        0.7
        ···etc
   10:00 - 11:00                 27                        9.0
   11:00 - 12:00                 28                        9.3
        ···etc
   23:00 - 0:00                   8                        2.7
Maximum logins 28 occurred between 11:00 and 12:00
Minimum logins 1  occurred between 0:00 and 1:00
```

7. The following two arrays represent the fixed and variable costs involved in producing each of eight items:

```
float fixed[]    = { 11.31, 12.12, 13.67, 11.91, 12.30,
                     11.8, 11.00, 12.00 } ;

float variable[] = { 1.12, 1.13, 3.14, 1.35, 2.20, 1.28,
                     1.00, 2.10 } ;
```

Write a program to input an item number in the range 1 to 8 along with the number of units produced. The program should then display the cost of producing that number of units.

8. Use two `for` loops to set all the diagonal elements of a 9 by 9 integer array to 1 and all the elements not on a diagonal to 0.
9. Write a program to input values to a 4 by 5 array, search the array for values that are less than 0 and display these values along with their row and column indices.
10. Write a program to input ten integer values into an array `unsorted`. Your program should then loop through `unsorted` ten times, selecting the lowest value during each pass. For each pass through the loop, the element in `unsorted` containing the lowest value is replaced with a large value (e.g.9999) after copying it into the next available element of another integer array `sorted`.
    This is illustrated below:

    `unsorted` at the start: 14 22 67 31 89 11 42 35 65 49
    `Sorted` at the start:

    `unsorted` after the first pass: 14 22 67 31 89 9999 42 35 65 49
    `sorted` after the first pass: 11

    `unsorted` after the second pass: 9999 22 67 31 89 9999 42 35 65 49
    `sorted` after the second pass: 11 14

    etc.

    Display the values in `sorted`. (Hint: see program P7C to determine the smallest value.)
11. In a magic square the rows, columns and diagonals all have the same sum. For example:

17	24	1	8	15
23	5	7	14	16
4	6	13	20	22
10	12	19	21	3
11	18	25	2	9

and

4	9	2
3	5	7
8	1	6

Write a program to read in a two-dimensional integer array from the keyboard and check if it is a magic square.

Chapter Eight
Pointers
第 8 章 指 针

8.1 Variable addresses（变量的地址）

Every variable used in a C program is stored in a specific place in memory. Each location in memory has a unique address, in the same way that every house in a street has a unique address.

C 程序中的每个变量都存储在特定的内存单元中，而每个内存单元都有唯一的地址，就如同街道中的每个房子都有唯一的地址一样。

You have already used the address operator & in the scanf() function. The address operator & is used to get the address of a variable. The next program uses & to get the address of two variables and display them on the screen.

在前面的scanf( )函数中，我们已经使用过取地址运算符&，取地址运算符&用于获得变量的地址。下面程序使用取地址运算符&得到变量的地址，然后将其显示在屏幕上。

Program Example P8A

```
1  /* Program Example P8A
2     To display the addresses of two variables. */
3  #include <stdio.h>
4  int main()
5  {
6    int var1 ;
7    char var2 ;
8
9    var1 = 1 ;
10   var2 = 'A' ;
11
12   printf( "var1 has a value of %d and is stored at %p\n",
13             var1, &var1 ) ;
14   printf( "var2 has a value of %c and is stored at %p\n",
15             var2, &var2 ) ;
16   return 0 ;
17 }
```

A sample run of this program is:

```
var1 has a value of 1 and is stored at 0xbfeb9438
var2 has a value of A and is stored at 0xbfeb943f
```

Figure 8.1 shows the variables var1 and var2 are stored in memory:

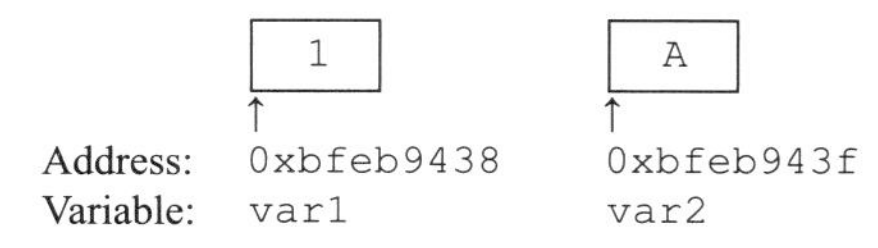

Figure 8.1 variables var1 and var2 and their hexadecimal addresses

如果读者的计算机给出不同的地址值，那么不必大惊小

Do not be surprised if your computer gives different addresses from the ones above. This is because various computers and operating systems will store variables at different memory locations. The above addresses start with 0x to indicate they are in hexadecimal (base 16). Depending on the compiler, the 0x may or may not be displayed.

怪，因为不同的计算机和操作系统会将变量存储在不同的内存单元中。上面的地址值以0x开头表示它是十六进制的（以16为基）。0x是否显示取决于所使用的编译器。

8.2 Pointer variables（指针变量）

A *pointer* variable is a variable that holds the address of another variable. A pointer variable is defined as follows:

指针变量是用于保存其他变量地址值的变量。

```
type *variable_name ;
```

where type is any data type (such as char, int, float, double, and so on) and variable_name is any valid C variable name. For example:

```
char *char_ptr; /* char_ptr is a pointer to a char variable. */
int  *int_ptr;  /* int_ptr  is a pointer to an int variable. */
```

Pointer definitions are read “backwards”, i.e. char* char_ptr means that char_ptr is a pointer to a char, and int *int_ptr means that int_ptr is a pointer to an int.
(Note that the * can be positioned anywhere between the data type and the variable name. Thus, int *int_ptr and int* int_ptr are equivalent.)
The next program defines and uses two pointers.

可以从后往前读指针变量的定义，例如，可以将char*char_ptr读做：char_ptr是一个指针变量，它指向一个字符型变量。int *int_ptr可以读做：int_ptr是一个指针变量，它指向一个整型变量。（注意，* 可以放在数据类型和变量名之间的任何位置，因此int *int_ptr与int* int_ptr是等价的。）

Program Example P8B

```
1  /* Program Example P8B
2     Demonstration of pointer variables. */
3  #include <stdio.h>
4  int main()
5  {
6    int var1 ;
7    char var2 ;
8    int  *ptr1 ;
9    char *ptr2 ;
10
11   var1 = 1 ;
12   var2 = 'A' ;
13
14   ptr1 = &var1 ; /* ptr1 contains the address of var1 */
15   ptr2 = &var2 ; /* ptr2 contains the address of var2 */
16
17   printf( "ptr1 contains %p\n", ptr1 ) ;
18   printf( "ptr2 contains %p\n", ptr2 ) ;
19   return 0 ;
20 }
```

Running this program displays the following:

```
ptr1 contains 0xbfe26d10
ptr2 contains 0xbfe26d1f
```

Figure 8.2 shows the program variables are stored in memory:

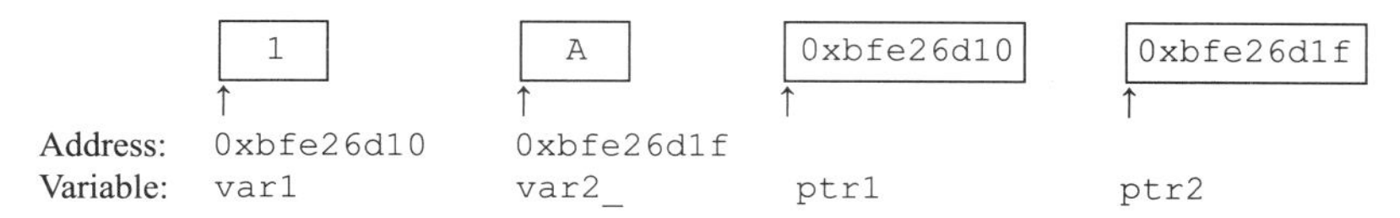

Figure 8.2 variables var1, var2, ptr1 and ptr2

The two variables ptr1 and ptr2 are used to store the addresses of the other two variables, var1 and var2.

Note: Depending on your computer and operating system, the values of ptr1 and ptr2 may be different.

8.3 The dereference operator *（解引用运算符 *）

The *dereference* operator * is used to access the value of a variable, whose address is stored in a pointer. For instance, *ptr means the value of the variable at the address stored in the pointer variable ptr.

The dereference operator * is also called the *indirection* operator.

解引用运算符*用于访问指针变量指向的变量的值，指针变量中保存的是该变量的地址。例如*ptr的值为在指针变量ptr存储的内存地址中的数据值。

解引用运算符*，也称为**间接寻址**运算符。

Program Example P8C

```
1  /* Program Example P8C
2     Demonstration of the dereference operator * */
3  #include <stdio.h>
4  int main()
5  {
6    int  var ;
7    int *ptr ;
8
9    var = 1 ;
10   ptr = &var ; /* ptr contains the address of var. */
11
12   printf( "ptr contains %p\n", ptr ) ;
13   printf( "*ptr contains %d\n", *ptr) ;
14   return 0 ;
15 }
```

This program displays two lines similar to the following:

```
ptr contains 0xbffc3548
*ptr contains 1
```

Note: Depending on your computer and operating system, the address at the end of the first line may be different.

The asterisk (*) is used in two different contexts in the above program. In line 7, the * is used to define ptr as a pointer to an int. In line 13, the * is used to access the value of the memory location, the address of which is in ptr. The two uses of * are not related.

在上面的程序中，星号（*）用在不同的上下文中。在程序第7行中，*用于将变量ptr定义为整型指针，而在第13行中，*用于访问ptr中存储的地址所对应的存储单元中的值，*的这两种用法之间没有任何关系。

程序的第10行将变量var的地址值赋值给指针变量ptr。

Line 10 of this program assigns the address of the variable `var` to the pointer variable `ptr`.

Line 12 displays the address contained in the pointer `ptr`. Note the use of the format specifier `%p` for displaying the pointer `ptr`'s value.

Line 13 displays the value at the address held in `ptr` by using the dereference operator *. This is called *dereferencing* the pointer `ptr`.

第12行将指针变量ptr中存储的地址值打印出来。注意，格式转换说明符%p用于输出指针变量的值。

程序的第13行使用解引用运算符*将指针变量ptr存储的内存地址中的数据值在屏幕上显示出来，这称为对指针ptr进行**解引用**。

8.4 Why use pointers?（为什么使用指针）

In program P8C above, the value of `*ptr` is the same as the value of `var`. So why go to all this trouble using a pointer when you can just as easily display the value of `var`?

The answer is that the above program is only used for demonstration purposes and more advanced programs use pointers in much more useful ways. For example, pointers allow different sections of a program to share the same data (see Chapter Eleven). Pointers are also used in complex data structures such as linked lists and binary trees.

指针允许程序的不同部分共享相同的数据（见第11章），指针还可以用于实现复杂的数据结构，例如链表和二叉树。

Programming pitfalls

1. Consider the following code segment:

```
int a = 1, b = 2, c ;
int *pa = &a, *pb = &b ;
c = *pa/*pb ;
```

The /* in the above assignment is interpreted as the start of a comment.
Use parentheses, as in:

```
c = (*pa)/(*pb) ;
```

or use spaces, as in:

```
c = *pa / *pb ;
```

2. If you are defining two or more pointers do so with

```
int *p1, *p2 ;
```

It is a common error to write

```
int *p1, p2 ;
```

In both cases `p1` is a pointer to an integer. In the first case `p2` is also a pointer to an integer, but in the second case `p2` is an integer.

Quick syntax reference

	Syntax	Examples
Defining a pointer	`type *variable ;` `(or type* variable ;)`	`int *pa ;` `float *pb, *pc ;`
Address	`&variable ;`	`int a ;` `pa = &a ;`
Dereference *	`*variable ;`	`*pa ;`

Exercises

1. Write a program to define the following variables and to display their addresses:

```
char c = 'a' ;
short int s = 1 ;
int i = 123456 ;
float f = 125.5 ;
double d = 1234.25 ;
```

Draw a diagram to illustrate the memory layout for these variables. How many bytes of memory are allocated for each of these variables?

2. Given the following:

```
int *p ;
int i, j ;
i = 40 ;
j = i ;
p = &i ;
```

Which of the following statements will change the value of `i` to `60`?

(a) `p = 60;` (b) `&i = 60;` (c) `*i = 60;` (d) `*p = 60;`

(e) `p = i + 20;` (f) `i = p + 20;` (g) `i = *(p + 20);` (h) `i = *p + 20;`

3. Given the following:

```
int   *i_ptr ;
float *f_ptr;
int i = 1, k = 2 ;
float f = 10.0 ;
```

which of these statements are valid?

(a) `i_ptr = &i;` (b) `f_ptr = &f;` (c) `f_ptr = f;`

(d) `f_ptr = &i;` (e) `k = *i;` (f) `k = *i_ptr;`

(g) `i_ptr = &k;` (h) `*i_ptr = 5;` (i) `i_ptr = &5;`

4. What does this program segment display?

```
int a, b ;
int *p1, *p2 ;

a = 1 ;
b = 2 ;
p1 = &a ;
p2 = &b ;
b = *p1 ;

printf( "\n%d %d\n", a, b ) ;
printf( "\n%d %d\n", *p1, *p2 ) ;

*p1 = 15 ;
printf( "\n%d %d\n", a, b ) ;

*p1 -= 3 ;
printf( "\n%d %d\n", a, b ) ;

*p2 = *p1 ;
printf( "\n%d %d\n", a, b ) ;

(*p1)++ ;
printf( "\n%d %d\n", a, *p2 ) ;

p1 = p2 ;
*p1 = 50 ;

printf( "\n%d %d\n", a, b ) ;
```

Chapter Nine
Pointers and Arrays
第 9 章　指针和数组

9.1 Pointers and one-dimensional arrays（指针和一维数组）

Pointers and arrays are directly related to one another. In C, the name of an array is equivalent to the address of the first element of the array; that is, the name of an array is a pointer to the first element of the array. Suppose you have the following array definition:

指针和数组之间是直接相关的，在C语言中，数组名代表数组的首地址，也就是说，数组名是指向数组第一个元素的指针。

```
int a[5] ;
```

The elements of this array are: a[0], a[1], a[2], a[3], and a[4]. The name of the array is a, and this is equivalent to the address of the first element; in other words, a is the same as &a[0]. The following program shows this.

Program Example P9A

```
1  /* Program Example P9A
2     This program shows that the name of an array is the same
3     as the address of its first element.                    */
4  #include <stdio.h>
5  #define NO_OF_ELS 5
6  int main()
7  {
8    int a[NO_OF_ELS] ;
9
10   printf( "a is %p and &a[0] is %p\n", a, &a[0] ) ;
11   return 0 ;
12 }
```

A sample run of this program is:

```
a is 0xbf9d35ec and &a[0] is 0xbf9d35ec
```

The actual addresses may be different on your system, but the two addresses will be the same nonetheless. The addresses are in hexadecimal (base 16), indicated by 0x at the start of the address.

Just as a is the address of the first element, a + 1 is the address of the second element, a + 2 is the address of the third element, and so on.

As the name of an array is a pointer to the first element of the array, the

也许该程序在你的计算机上显示的实际地址与此不同，但是a与&a[0]这两个地址值一定是相同的。地址值是十六进制的（以16为基数），因此用以0x开头的地址值来表示。a代表了数组的首地址，与之类似，a+1表示数组第2个元素的地址，a+2表示数组第3

indirection operator * can be used to access the elements of the array. The next program demonstrates the technique by displaying the elements of an array using the indirection operator *.

个元素的地址，以此类推。由于a是指向数组第一个元素的指针，因此可以使用间接寻址运算符*来访问数组元素。

Program Example P9B

```
1  /* Program Example P9B
2     To demonstrate accessing the elements of an array using
3     element addresses rather than subscripts.            */
4  #include <stdio.h>
5  #define NO_OF_ELS 5
6  int main()
7  {
8    int a[NO_OF_ELS] = { 10, 13, 15, 11, 6 } ;
9    int i ;
10
11   for ( i = 0 ; i < NO_OF_ELS ; i++ )
12     printf( "Element %d is %d\n", i, *(a+i) ) ;
13   return 0 ;
14 }
```

This program displays the elements of the array as follows:

```
Element 0 is 10
Element 1 is 13
Element 2 is 15
Element 3 is 11
Element 4 is 6
```

If you changed `*(a+i)` to `a[i]` in line 12, the program would produce the same output. Thus:

`*(a+0)` or `*a` is equivalent to `a[0]`
`*(a+1)` is equivalent to `a[1]`
`*(a+2)` is equivalent to `a[2]`, and so on.

The parentheses in the expression `*(a+i)` are important. Without them the expression `*a+i` would add the first element of the array `a` and `i` together. You can use pointers to access the elements of any array, not just an array of integers. Suppose you have an array of `floats` defined as

表达式*(a+i)中的括号非常重要，去掉括号的表达式*a+i的值为数组a的第一个元素的值加i。使用指针可以访问任意类型的数组元素，不只限于整型数组。

```
float numbers[100] ;
```

then `numbers[i]` is equivalent to `*(numbers+i)`.

Although the name of an array is a pointer to the first element of the array, you cannot change its value, because it is a constant pointer. Thus, expressions such as `a++` or `numbers+=2` are invalid, because both `a` and `numbers` are array names. You can, however, assign the name of an array to a pointer variable of the same type. For example:

虽然数组名是指向数组首元素的指针，但是由于数组名是一个常量指针，因此它的值是不能改变的。由于a和numbers都是数组名，所以类似于a++或numbers+=2这样的表达式是非法的。不过，可以将数组名赋值给相同类型的指针变量。

```
int a[5], *p ;

p = a ;      /* Valid: assignment of a constant to a variable.   */
a++ ;        /* Invalid: the value of a constant cannot change.  */
p++ ;        /* Valid: p is a variable. p now points to
                element 1 of the array a.                        */
p-- ;        /* Valid: p points to element 0 of the array a.     */
p += 10 ;    /* Valid, but p is outside the range of the array a,
                so *p is undefined. A common error.              */
p = a - 1 ; /* Valid, but p is outside the range of the array.  */
```

A constant may be added to or subtracted from the value of a pointer, allowing access to different memory locations. However, not all arithmetic operations are permissible on pointers. For example, the multiplication of two pointers is illegal, because the result would not be a valid memory address.

Why bother using pointers when you can use subscripts? One reason is that pointers are more efficient than subscripts. Pointers usually result in less code being generated by the compiler and are slightly faster in execution.

可以通过让指针变量的值加上或者减去一个常量值来访问不同的内存单元。但是，并非所有的算术运算都能应用于指针。例如，两个指针变量的乘法运算是非法的，这是因为，乘法运算的结果可能不是一个有效的内存地址值。

既然可以使用下标来访问数组元素的值，那么为什么还要费力去使用指针呢？其中的一个原因是使用指针比使用下标效率高，使用指针编写的程序编译生成的代码比使用数组编写的程序编译生成的代码少，其执行速度也稍快。

9.2 Pointers and multi-dimensional arrays（指针和多维数组）

As with one-dimensional arrays, you can access the elements of a multi-dimensional array using pointers. However, as the number of dimensions of an array increases, the pointer notation becomes increasingly complex. Suppose you have the following definition:

与一维数组类似，也可以使用指针来访问多维数组的元素。但是，随着数组维数的增加，指针标记也变得更为复杂。

```
int a[3][2] = { { 4, 6 },
                { 1, 3 },
                { 9, 7 } } ;
```

C stores this two-dimensional array as an “array of arrays”. This means that a is a one-dimensional array whose elements are themselves one-dimensional arrays of integers.

As with a one-dimensional array, the name of the array is a pointer to the first element of the array. Therefore a is equivalent to &a[0]. a[0] is itself an array of two integers, which means that a[0] is equivalent to &a[0][0].

C语言将二维数组存储为“数组中的数组”。例如，a本身是一个一维整型数组，而它的元素本身又是一维整型数组。在一维数组中，数组名是指向数组第一个元素的指针，因此a等价于&a[0]，而a[0]本身又是包含两个整型元素的数组，这就意味着a[0]等价于&a[0][0]。

a →	a[0]→	4	6
	a[1]→	1	3
	a[2]→	9	7

a[0], a[1] and a[2] are pointers (data type is int*) and a is a pointer to a pointer (data type is int**).

a[0]、a[1]、a[2]均为指针，数据类型为int *；a是指向指针的指针，数据类型为int **。

`a[0]` is the address of the first element in the first row of the array.
`*a[0]` is `a[0][0]`, which is `4`.

`a[1]` is the address of the first element in the second row.
`*a[1]` is `a[1][0]`, which is `1`.

`a[2]` is the address of the first element in the third row.
`*a[2]` is `a[2][0]`, which is `9`.

`a[0]+1` is the address of the second element in the first row.
`*(a[0]+1)` is `a[0][1]`, which is `6`.

`a[1]+1` is the address of the second element in the second row.
`*(a[1]+1)` is `a[1][1]`, which is `3`.

`a[2]+1` is the address of the second element in the third row.
`*(a[2]+1)` is `a[2][1]`, which is `7`.

Using the fact that

`*a` is the same as `a[0]`
and `*(a+1)` is the same as `a[1]`
and `*(a+2)` is the same as `a[2]`

you can conclude the following:

1. `a[0][0]` is `*a[0]` is `*(*a)` or `**a`
2. `a[1][0]` is `*a[1]` is `*(*(a+1))`
3. `a[2][0]` is `*a[2]` is `*(*(a+2))`
4. `a[0][1]` is `*(a[0]+1)` is `*(*a+1)`
5. `a[1][1]` is `*(a[1]+1)` is `*(*(a+1)+1)`
6. `a[2][1]` is `*(a[2]+1)` is `*(*(a+2)+1)`

9.3 Dynamic memory allocation（动态内存分配）

Up to this point, when you define an array in C you must specify in advance the number of elements in the array before the program is compiled and executed. You may then find that either the program does not use all the elements you have specified or it requires more. To overcome this, what is required is the ability of the program to specify the exact number of elements in an array while it is executing. This can be done using the standard library memory allocation functions.

到目前为止，在C程序中定义数组时，必须在程序编译运行前预先指定数组元素的个数。可能会发现，要么定义的数组元素太多，程序不会用到数组的所有元素，要么就是定义的数组元素太少，数组元素不够用。为了解决这个问题，需要使程序具有在运行过程中指定准确的元素个数的能力，这种功能可以通过标准函数库中的内存分配函数来实现。

9.3.1 The `malloc()` function

This function allocates a contiguous block of memory and returns a pointer to the start of the allocated block. The general format of **malloc( )** (memory allocation) is:

函数malloc( )用于分配一段连续的内存空间，并返回指向该内存空间首地址的指针。

```
pointer = malloc( size ) ;
```

where `pointer` is a pointer (an address) to the start of the allocated memory block and `size` is the number of bytes in the block. For example:

```
pointer = malloc( 40 ) ; /* Allocate 40 bytes of memory. */
```

If the block of memory cannot be allocated, `malloc()` returns `NULL` to `pointer`.

如果内存空间分配失败，那么函数malloc( )给pointer返回空指针NULL。

The next program uses `malloc()` to allow the user to specify the number of elements in an integer array.

Program Example P9C

```
1  /* Program Example P9C
2     To demonstrate dynamic memory allocation using malloc(). */
3  #include <stdio.h>
4  #include <stdlib.h>
5  int main()
6  {
7    int *int_array ;
8    int no_els, no_bytes, i ;
9
10   printf( "Enter the number of elements: " ) ;
11   scanf( "%d", &no_els ) ;
12
13   /* Calculate the number of bytes required by the array. */
14   no_bytes = no_els * sizeof( int ) ;
15   /* Allocate the memory for the array. */
16   int_array = (int *) malloc( no_bytes ) ;
17   if ( int_array == NULL )
18   {
19     printf( "Cannot allocate memory\n" ) ;
20     return EXIT_FAILURE ;
21   }
22
23   /* Enter a value for each element of the array. */
24   for ( i=0 ; i< no_els ; i++ )
25   {
26     printf( "Enter element %d: ", i ) ;
27     scanf( "%d", &int_array[i] ) ;
28   }
29
30   /* Display the element values just entered. */
31   for ( i=0 ; i < no_els ; i++ )
32     printf( "Element %d is %d\n", i, int_array[i] ) ;
33
34   free( int_array ) ; /* Free the allocated memory. */
35   return 0 ;
36 }
```

Here is a sample run of this program:

```
Enter the number of elements: 4
Enter element 0: 23
```

```
Enter element 1: 11
Enter element 2: 7
Enter element 3: 88
Element 0 is 23
Element 1 is 11
Element 2 is 7
Element 3 is 88
```

Line 4 is a new `#include` preprocessor directive. This line is necessary when any of the memory allocation functions are used in a program.

程序第4行是一个新的#include编译预处理指令，当程序中使用内存分配函数时，必须使用该指令。

Line 11 gets the number of elements in the array from the user.

第11行的功能是读取用户输入的数组元素个数。

The number of bytes required for an integer array is the number of elements in the array multiplied by the number of bytes required to store an integer. This value is calculated on line 14 and used by `malloc()` on line 16.

整型数组占用内存空间的字节数等于数组的元素个数乘以整型占用的内存字节数，这个值在程序的第14行进行计算，在第16行语句的malloc( )函数中使用。malloc( )函数前面的强制类型转换(int *)将函数malloc( )的返回值强转为int_array的数据类型，即(int *)。

The type cast `(int *)` preceding `malloc()` changes the pointer returned by `malloc()` into the data type of `int_array`, i.e. `int *`.

If the memory is not available, the pointer `int_array` will be `NULL` and the error message on line 19 is displayed.

如果可用内存不足，那么指针变量int_array的值将为NULL，程序第19行显示错误提示信息。

The `return` statement on line 20 terminates the program if the requested memory cannot be allocated. `EXIT_FAILURE` (defined in `stdlib.h` included on line 4) is an *exit status code* that is passed back to the operating system to indicate failure. (Note: `EXIT_SUCCESS` can be used in place of `0` on line 35.)

如果申请内存没有成功，那么将由第20行的return语句结束程序的执行。EXIT_FAILURE（在stdlib.h 中定义，由第4行的编译预处理指令将其包含到文件中）是返回操作系统表示失败的**退出状态码**。（注意，第35行return语句中的0也可以用EXIT_SUCCESS代替）。

If the memory has been successfully allocated, `int_array` will point to (i.e. hold the address of) the start of the memory block.

如果内存分配成功，那么指针int_array将指向已分配内存块的首地址。

Lines 24 to 28 input the values for the array, and lines 31 to 32 display the values. Note that subscripts are used to access each element of the array, but pointer notation could also be used. Using pointer notation, line 27 becomes

程序的第24行~第28行输入数组元素的值，第31行~第32行输出数组元素的值。注意，这里使用下标法来访问数组元素值，也可以使用指针标记法来访问数组元素值。

```
scanf( "%d", int_array + i ) ;
```

and line 32 becomes

```
printf( "Element %d is %d\n", i, *(int_array + i) ) ;
```

Finally, the allocated memory is freed using the **free()** function. This is not necessary here, as the program is finishing and the memory will be freed by the operating system. However, `free()` could be used at any point within a program to free a block of previously allocated memory that was no longer required.

在程序的最后，使用函数free( )释放已分配的内存空间，由于它是程序的最后一行，操作系统在程序结束时会自动释放程序占用的所有内存，因此本程序在这里也可以不用free( )函数释放已分配的内存。但是，若要在程序的中间位置释放由系统之前分配的不再使用的内存，那么就一定要使用free( )函数去释放它。

9.3.2 The `calloc()` function

The **calloc()** function, like `malloc()`, allocates a contiguous block of memory and returns a pointer to the start of the allocated block. However, unlike `malloc()`, `calloc()` also initialises the block of memory with zero values. The general format of `calloc()` (meaning calculated allocation) is:

与函数malloc( )类似，函数calloc( )也用于分配一块连续的内存空间，然后返回该内存空间首地址的指针。但是，与函数malloc( )不同的是，函数calloc( )将其分配的所有内存单元都初始化为0。

```
pointer = calloc( number, size ) ;
```

where `pointer` is a pointer to the allocated memory, `number` is the number of elements for which memory is to be allocated, and `size` is the size of each element.

pointer为指向分配内存空间首地址的指针，number为要分配内存空间包含的元素个数，size为每个元素占用内存空间的大小。

If the block of memory cannot be allocated, `calloc()` returns `NULL` to `pointer`. For example:

如果内存分配失败，则函数calloc( )返回空指针NULL给pointer。

```
/* Allocate 10 elements of 4 bytes each (a total of 40 bytes)
   and zeroise each element. */
```

```
pointer = calloc( 10, 4 ) ;
```

The next program uses `calloc()` to dynamically allocate and initialise a floating-point array with zero values.

Program Example P9D

```
1  /* Program Example P9D
2     To demonstrate dynamic memory allocation using calloc(). */
3  #include <stdio.h>
4  #include <stdlib.h>
5  int main()
6  {
7    float *float_array ;
8    int no_els, i ;
9
10   printf( "Enter the number of elements: " ) ;
11   scanf( "%d", &no_els ) ;
12
13   /* Allocate no_els each of size sizeof(float) and
14      initialise all elements to 0 */
15   float_array = (float *) calloc( no_els, sizeof(float) ) ;
16   /* Note difference between calloc() and malloc(). */
17   if ( float_array == NULL )
18   {
19     printf( "Cannot allocate memory\n" ) ;
20     return EXIT_FAILURE ;
21   }
22   /* Display the contents of the memory just allocated. */
23   for ( i = 0; i < no_els; i++ )
24     printf( "Element %d is %.1f\n", i, float_array[i] ) ;
25
26   free( float_array ) ; /* Free the allocated memory. */
```

```
27   return 0 ;
28 }
```

In this program the user specifies the number of elements in the array. `calloc()` then calculates the number of bytes required and allocates and initialises the required memory. Lines 23 to 24 display the values in the array. All these values will be 0.

Here is a sample run of this program:

```
Enter the number of elements: 5
Element 0 is 0.0
Element 1 is 0.0
Element 2 is 0.0
Element 3 is 0.0
Element 4 is 0.0
```

9.3.3 The `realloc()` function

The **realloc( )** function changes the size of a previously allocated block of memory. The general format of `realloc()` (meaning re-allocate) is:

```
ptr_new = realloc( ptr_old, new_size ) ;
```

where `ptr_old` is a pointer to a previously allocated memory block and `new_size` is the new memory size. If it is not possible to assign the required new memory size, `ptr_new` will be `NULL`. Note that, depending on the availability of memory, `ptr_new` may either contain the same address as `ptr_old` or contain a new address.

The next program dynamically allocates memory by continuously extending the memory for each new element value entered by the user. In this program the number of elements is not specified by the user in advance of the memory allocation.

使用realloc( )函数可以更改先前已分配的内存块的大小。ptr_old为指向先前已分配的内存块的指针，new_size为新分配的内存块的大小。如果系统不能按指定的大小分配新的存储空间，那么函数将返回空指针NULL给ptr_new。注意，根据可用于分配的内存的具体情况，ptr_new得到的地址值既可能与ptr_old相同，也可能不同，即一个新的地址值。

下面的程序在用户每输入一个新的元素值时连续地扩展动态分配的内存，程序在内存分配之前，用户未事先指定元素的个数。

Program Example P9E

```
1  /* Program Example P9E
2     To demonstration dynamic memory allocation using realloc(). */
3  #include <stdio.h>
4  #include <stdlib.h>
5  int main()
6  {
7    char reply ;
8    int no_els = 0, i, el_size ;
9    float *float_array ;
10
11   el_size = sizeof( float ) ; /* Store the size of one element. */
12   /* Start the array with a memory allocation for one element. */
13   float_array = (float *) malloc( el_size ) ;
14   if ( float_array == NULL )
15   {
```

```
16      printf( "Cannot allocate memory\n" ) ;
17      return EXIT_FAILURE ;
18    }
19    do
20    {
21      printf( "\nEnter a numeric value: " ) ;
22      scanf( "%f", float_array+no_els ) ;
23      no_els++ ;
24      printf( "Any more values? (y or n) " ) ;
25      scanf( "%1s", &reply ) ;
26      if ( reply == 'Y' || reply == 'y' )
27      {
28        /* Enlarge the array by the size of one element. */
29        float_array =
30              (float *) realloc( float_array,el_size*(no_els+1) ) ;
31        if ( float_array == NULL )
32        {
33           printf( "Cannot allocate any more memory\n" ) ;
34           return EXIT_FAILURE ;
35        }
36      }
37    }
38    while ( reply == 'Y' || reply == 'y' ) ;
39
40    /* Display the values just entered. */
41      for ( i=0; i<no_els; i++ )
42        printf( "\nElement %d is %f\n", i, float_array[i] ) ;
43
44      free( float_array ) ; /* Free the allocated memory. */
45      return 0 ;
46 }
```

A sample run of this program follows.

```
Enter a numeric value: 8
Any more values? (y or n) y

Enter a numeric value: 4
Any more values? (y or n) y

Enter a numeric value: 2
Any more values? (y or n) n

Element 0 is 8.000000

Element 1 is 4.000000

Element 2 is 2.000000
```

9.3.4 Allocating memory for multi-dimensional arrays

In C, multi-dimensional arrays are implemented as "arrays of arrays". To fully understand dynamic memory allocation for multi-dimensional arrays, you need to be familiar with section 9.2.

在C语言中，多维数组被当做是“数组中的数组”，只有熟悉了9.2节的内容，才能充分

The next program demonstrates the dynamic allocation of a two-dimensional array of integers.

理解多维数组的动态内存分配问题。

Program Example P9F

```
1  /* Program Example P9F
2     To demonstrate dynamic memory allocation of a
3     two-dimensional array using malloc().        */
4  #include<stdio.h>
5  #include<stdlib.h>
6  int main()
7  {
8    int no_rows, no_columns, i, j ;
9    int **a ; /* a is the name of the array. */
10
11   /* Get the number of rows and columns in the array. */
12   printf( "Enter the number of rows: " ) ;
13   scanf( "%d", & no_rows ) ;
14   printf( "Enter the number of columns: " ) ;
15   scanf( "%d", & no_columns ) ;
16
17   /* Allocate memory for each element of a.
18      Each element of a is a pointer to an array of integers. */
19   a = ( int ** ) malloc( no_rows * sizeof(int *) ) ;
20
21   /* Allocate memory for each array (row) in a. */
22   for( i = 0 ; i < no_rows ; i++ )
23   {
24     a[i] = ( int * ) malloc( no_columns * sizeof(int) ) ;
25     if ( a[i]== NULL )
26     {
27       printf( "Cannot allocate memory\n" ) ;
28       return EXIT_FAILURE ;
29     }
30   }
31
32   /* Enter a value for each element of the array, row by row. */
33   for ( i = 0 ; i < no_rows ; i++ )
34   {
35     for ( j=0 ; j < no_columns; j++ )
36     {
37       printf( "Enter element [%d,%d]: ", i, j ) ;
38       scanf( "%d", &a[i][j] ) ;
39     }
40   }
41
42   /* Display the element values just entered. */
43   printf( "\nThe values in the array are:\n" ) ;
44   for ( i = 0 ; i < no_rows; i++ )
45   {
46     for ( j = 0 ; j < no_columns ; j++ )
```

```
47      {
48        printf( "Element [%d,%d] is %d\n", i, j, a[i][j] ) ;
49      }
50    }
51
52    for( i = 0 ; i < no_rows ; i++ )
53      free( a[i] ) ;
54    free( a ) ;
55    return 0 ;
56 }
```

A sample run of this program is:

```
Enter the number of rows: 2
Enter the number of columns: 3
Enter element [0,0]: 6
Enter element [0,1]: 11
Enter element [0,2]: 21
Enter element [1,0]: -8
Enter element [1,1]: 121
Enter element [1,2]: 32

The values in the array are:
Element [0,0] is 6
Element [0,1] is 11
Element [0,2] is 21
Element [1,0] is -8
Element [1,1] is 121
Element [1,2] is 32
```

The `return` statement on line 28 terminates the program if the requested memory cannot be allocated.

Lines 54 and 53 free the memory allocated in lines 19 and 24. For each pointer returned from `malloc()` in lines 19 and 24 there is a corresponding call to `free()` with that pointer.

如果分配内存失败，则第28行的return语句结束程序的执行。EXIT_FAILURE（在第5行包含的头文件stdlib.h中定义）是返回给操作系统表示失败的退出状态码。注意第55行中的0也可以用EXIT_ SUCCESS代替。

程序第54行和第53行分别将第19行和第24行分配的内存空间释放。对于由第19行和第24行的函数malloc( )返回的每一个指针，都要有一个free( )函数调用与之相对应。

Programming pitfalls

1. Pointers, like any other variable, are not automatically initialised in C. Do not use a pointer until it has been assigned a value. For example:

```
int *p ;
*p = 100 ;    /* Where is p pointing to? */
```

2. Do not try to access a non-existent array element. For example:

```
int a[10] ;
int i ;
for ( i = 1 ; i <= 10 ; i++ )
  a[i] = 0 ;
```

The loop in this program segment starts at `1` and stops at `11`, accessing `a[1],a[2],a[3] ... a[10]`. This is incorrect, because the elements of `a` are `a[0],a[1],a[2], ... a[9]`.

The `for` loop should be:

```
for ( i = 0 ; i < 10 ; i++ )
  a[i] = 0 ;
```

3. Always test the value of the pointer returned by `malloc()`, `calloc()`, and `realloc()`. If the value is `NULL`, the requested amount of memory is not available.

1. 与其他类型的变量一样，C语言中的指针变量是不会被自动初始化的，因此在使用指针变量之前，必须先对其进行初始化。
2. 不要试图去访问并不存在的数组元素。
在这个程序片段的循环语句中，通过控制循环变量i的值从1变化到10，试图访问元素a[1]，a[2]，a[3]，…，a[10]，而数组a的元素实际为a[0]，a[1]，a[2]，…，a[9]，因此这个程序是错误的。
3. 使用函数malloc( )、calloc( )与realloc( )时，一定要测试其返回的指针值，如果函数返回NULL，则说明内存申请失败。

Quick syntax reference

	Syntax	Example

The name of an array is a pointer to the first element of the array		`int a[10] ;`
		`/*`
		`a is the same as &a[0].`
		`*/`

Memory allocation	`#include <stdlib.h>`
	`ptr=(cast)malloc(mem_size);`
	`ptr=(cast)calloc(no_els,`
	`el_size);`
	`ptr=(cast)realloc(ptr,`
	`new_size);`
		`/* Allocate memory for`
		`ten integers. */`
		`int n ;`
		`int *p ;`
		`n=10*sizeof(int) ;`
		`p=(int *)malloc(n) ;`

Free memory	`free(ptr) ;`	`free(p);`

Exercises

1. What is wrong with each of the following?

(a)
```
float a[5] ;
int *p ;
p = a ;
```
(b)
```
float a[5] ;
float *p ;
p =&a ;
```
(c)
```
float a[5] ;
float *p ;
*p = a ;
```

2. What is the output from the following program segment?

```
char c[] = { 'G', 'D', 'K', 'K', 'N' } ;
int i ;
```

```
for ( i = 0 ; i < 5 ; i++ )
  printf( "%c", *( c + i ) + 1 ) ;
```

3. What does this program segment do?

```
int a[5] ;
int i ;
int *p ;

for ( i = 0 ; i < 5 ; i++ )
  scanf( "%d", a+i ) ;
for ( p = a ; p < a+5 ; p++ )
  printf( " %d ", *p ) ;
```

4. What is the value of `*p`, `*p+4` and `*(p+4)` in each of the following?

(a)
```
int one_d[] = { 1, 3, 4, 5, -1 } ;
int *p ;
p = one_d ;
```
(b)
```
char c[] = { 'A', 'a', 'B', 'b', 'C', 'c', 'D', 'd' } ;
char *p ;
p = c + 3 ;
```
(c)
```
float f[] = { 1.25, 11.0, 9.5, 3.5, 6.5, 1.0 } ;
float *p ;
p = f ;
```
(d)
```
int two_d[3][6] = {  { 1, 5, 0, 9, 11, -4 },
                     { 3, 9, 4, 6, 10, 123 },
                     { 11, 7, 4, -10, 19, 15 } } ;
int *p ;
p = two_d[1] ;
```

5. Given the following definitions:

```
int numbers[10] = { 1, 7, 8, 2 } ;
int *ptr = numbers ;
```

what is in the array `numbers` after each of the following?

(a) `*( ptr + 4 ) = 10 ;`

(b) `*ptr-- ;`

(c) `*( ptr + 3 ) = *( ptr + 9 ) ;`

(d) `ptr++ ;`

(e) `*ptr = 0 ;`

(f) `*( numbers + 1 ) = 1 ;`

6. Given the following variable definitions:

```
char letters[3] = { 'A', 'B', 'C' } ;
char *ptr = letters + 1 ;
char c ;
```

what is in the variable `c` after each of the following?

(a) `c = *( letters + 2 ) ;`

(b) `c = *( ptr + 1 ) ;`

(c) `c = *++ptr ;`

(d) `c = ++*ptr ;`

(e) `c = *ptr++ ;`

7. If `a` is a 6 by 8 array, what elements of `a` do the following expressions access?

   (a) `*a[2]`

   (b) `*( a[2] + 7 )`

   (c) `*(*a)`

   (d) `*(*( a + 5 ) + 2 )`

8. What is the difference between the memory allocation functions `malloc()` and `calloc()`? What may cause these functions to return `NULL`?

9. Using `malloc()` or `calloc()`, write a program to input a specified number of integer values into an array and to display the array and the sum of the elements in the array. Use pointers, not subscripts, in the program.

10. Given an array such as

```
char chars[] = { 'a', ' ', 'b', ' ', 'c', ' ', ' ', 'd' } ;
```

write a program that replaces all the blank elements in a character array with the underline character `'_'`. Use a pointer, rather than a subscript, to access the elements of the array.

11. Given the following arrays,

```
float litres[] = { 11.5, 11.21, 12.7, 12.6, 12.4 } ;
float miles[] = { 471.5, 358.72, 495.3, 453.6, 421.6 } ;
int mpl[5] ; /* Miles per litre. */
```

write a program to calculate and display the value of each element of `mpl`. Use pointers, rather than subscripts, to access the elements of each array.

Chapter Ten
Strings
第 10 章　字　符　串

10.1　String literals（字符串）

A *string literal* (or *string*) is any sequence of characters enclosed in double quotation marks. For example, `"Hello"` and `"xyz123"` are string literals. We have used strings in `printf()` to display a message on the screen as in

字符串是用双引号括起来的任意一串字符。例如，"Hello"和"xyz123"都是字符串。前面我们曾在函数printf( )中使用字符串向屏幕输出信息。

```
printf( "Input a value" ) ;
```

or

```
printf( "The result is" ) ;
```

The characters of a string are stored in contiguous memory locations. A string is terminated by the null character `'\0'`. The null character is a single byte of memory containing a zero. For example, the string `"Hello"` will be stored in memory as shown in Figure 10.1 below.

字符串中的字符在内存中是连续存放的，以空字符'\0'作为字符串的结束符，'\0'占1个字节的内存空间，值为0。

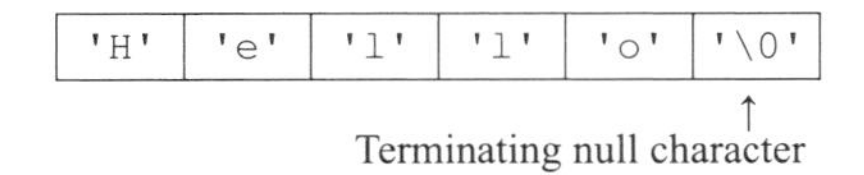

Figure 10.1　the string `Hello`

To use the double quotation mark (") or backslash (\) characters in a string you must precede them with a backslash (\). For example,

如果字符串中有双引号（"）或反斜线（\）字符，那么必须在该字符的前面再放置一个反斜线字符（\）对其进行转义。

```
printf( "\"I love C,\" said the student." ) ;
printf( "c:\\ is the root directory of drive c." ) ;
```

will display:

```
"I love C," said the student.
c:\ is the root directory of drive c.
```

The `\\` and `\"` are just two examples of the special character constants called *escape sequences*. An escape sequence consists of a backslash (\) followed by a letter or by a combination of octal or hexadecimal digits.

Escape sequences are typically used to specify actions such as backspace, newline, tab, sounding the bell, etc. Table 10.1 contains a list of C escape sequences.

\\和\"是两个特殊的字符常量，这样的字符称为**转义序列**。转义序列由反斜线（\）及其后面的字母或者八进制、十六进制的数字来组成。

转义序列的典型应用是实现退格、换行、制表、响铃等特殊的操作。

Table 10.1 escape sequences

Character	Meaning
\a	Alert (bell)
\n	Newline
\t	Tab
\b	Backspace
\r	Carriage return
\f	Form feed
\\	Backslash
\"	Double quotation mark
\'	Single quotation mark
\0	Null
\ddd	This ASCII character with the octal code ddd, where ddd is a 1 to 3 digit octal number, e.g. \251 is ©
\xdd	The ASCII character with the hexadecimal code dd, where dd is a 1 or 2 digit hexadecimal number, e.g. /xA9 is ©

Note: When an escape sequence appears in a string it counts as a single character.

注意：当转义序列出现在字符串中的时候，按单个字符计数。

10.2 Long character strings（长字符串）

What if a string is too long to fit in a single line of a C program? One way to type a string that exceeds the length of a line is to type a backslash, then the Enter key. For example, the string

在C程序中，如果字符串太长，一行写不下，该如何处理呢？一种解决办法是将其分为多行，除末行末尾是回车外，在其余每行的末尾添加反斜线。

```
"This string can be bro\
ken over two \
or more lines."
```

is identical to the string

```
"This string can be broken over two or more lines."
```

In addition, when two or more strings are separated only by whitespace characters, they are concatenated into a single string. For example,

另一种方法是，将一个长字符串分为两个或多个字符串，使用空白字符将它们分开，那么它们可以由系统自动连接为一个字符串。

```
printf( "This is the first half of a string,"
" and this is the second half." ) ;
```

will display:

```
This is the first half of a string, and this is the second half.
```

Provided each part of the string is enclosed in double quotation marks, the individual parts of the string will be concatenated into a single string.

只要字符串的每个部分都用双引号括起来，那么它们就可以被系统自动连接为一个字符串。

10.3 Strings and arrays（字符串和数组）

In C, a string is an array with elements of type `char`. The number of elements in the array is equal to the number of characters in the string, plus one for the terminating null character `'\0'`.

在C语言中，字符串就是元素为字符型的数组。由于在字符串的末尾有一个字符串结

As a string is an array of type `char`, it can be initialised in the same way as any other array. For example:

束符'\0'，所以数组元素的个数应等于字符串中实际字符的个数加1。

```
char greetings[] = { 'H', 'e', 'l', 'l', 'o', '\0' } ;
```

This statement initialises a six-element `char` array `greetings` with the character constants that spell the word `"Hello"`. Note that the last element of the array `greetings` is the null character (`'\0'`). Without the null character `greetings` is a character array but not a proper string. An easier way to initialise `greetings` is:

由于字符串就是一个字符型数组，所以字符串可以使用与其他字符数组同样的方式进行初始化。

```
char greetings[] = "Hello" ;
```

In the above definition of the array `greetings`, the compiler determined the size of the array by the number of letters in the array plus 1 (for the null character `'\0'`).

Just as with other array types, you can access the individual elements of `greetings` using subscripts or the indirection operator `*`:

与其他类型的数组一样，我们可以使用下标或者间接寻址运算符两种方式来访问数组greetings中的元素。

```
greetings[0] is 'H'
greetings[1] is 'e'
```

and

```
*greetings is 'H'
*( greetings +1 ) is 'e' and so on.
```

If you specify the size of the array and the string is shorter than this size, the remaining elements of the array are initialised with the null character `'\0'`. For example

如果指定的数组长度大于对其进行初始化的字符串的长度，那么编译器将数组中的多余元素全部初始化为空字符'\0'。

```
char greetings[9] = "Hello" ;
```

initialises `greetings` as shown in Figure 10.2 below.

'H'	'e'	'l'	'l'	'o'	'\0'	'\0'	'\0'	'\0'

Figure 10.2 the string `greetings`

If the array size is too small to hold all the characters of the string, some compilers (but not all) will generate an error message.

如果指定的数组长度太小，不能容纳下字符串中的所有字符，那么某些编译器（不是全部）将给出错误提示信息。

10.4 Displaying a string（显示一个字符串）

The next program shows how to display a string using `printf()` with `%s` in the format string.

Program Example P10A

```
1  /* Program Example P10A
2     To demonstrate the use of %s format string in printf(). */
3  #include <stdio.h>
4  int main()
5  {
```

```
6     char love_note[9] = "I love C" ;
7
8     /* Display column numbers. */
9     printf( "1234567890123456789012345\n" ) ;
10    printf( "%s", love_note) ;     /* No width specified in %s      */
11    printf( ".\n" ) ;              /* Display a . after the string */
12    printf( "%20s", love_note ) ; /* Width 20, right justified.   */
13    printf( ".\n" ) ;
14    printf( "%-20s", love_note ); /* Width 20, left justified.    */
15    printf( ".\n" ) ;
16    printf( "%.6s", love_note ) ; /* First 6 characters.          */
17    printf( ".\n" ) ;
18    printf( "%20.6s", love_note ) ; /* First 6 characters,        */
19    printf( ".\n" ) ;                /* width 20, right justified. */
20    printf( "%-20.6s", love_note ); /* First 6 characters,        */
21    printf( ".\n" ) ;                /* width 20, left justified.  */
22    return 0 ;
23 }
```

This program will display the following lines:

```
1234567890123456789012345
I love C.
            I love C.
I love C            .
I love.
              I love.
I love
```

Line 9 of program P10A displays a line of numbers so that you can see the columns that the characters of the string `love_note` are displayed in.

Lines 10, 12, 14, 16, 18 and 20 display the string `love_note` in various formats. Lines 11, 13, 15, 17, 19 and 21 display a full stop at the end of a line so that you can see how many columns were used by each format in the `printf()` functions.

From the output you can see the following:

`%s`	displays the string exactly as it is (line 10)	按照字符串本身的格式输出（对应程序的第10行）
`%20s`	displays the string right-aligned in a field of width 20 (line 12)	以右对齐方式、以20为域宽，输出字符串（对应程序的第12行）
`%-20s`	displays the string left-aligned in a field of width 20 (line 14)	以左对齐方式、以20为域宽，输出字符串（对应程序的第16行）
`%.6s`	displays the first six characters of a string (line 16)	输出字符串的前6个字符（对应程序的第16行）
`%20.6s`	displays the first six characters of a string in a field of width 20 (line 18)	以右对齐方式、以20为域宽，输出字符串的前6个字符（对应程序的第18行）
`%-20.6s`	displays the first six characters left-aligned in a field of width 20 (line 20)	以左对齐方式、以20为域宽，输出字符串的前6个字符（对应程序的第20行）

10.5 The puts() function [puts( ) 函数]

puts( ) is another C function used to display a string. In addition to displaying a string, puts() also places the cursor at the start of the next line on the screen. You could replace line 9 of program P10A with the statement

C语言中另一个用于输出字符串的函数是puts( )，除了向屏幕输出字符串之外，puts( )函数还要将光标移至下一行的起始位置。

```
puts( "1234567890123456789012345" ) ;
```

Note that there is no need for the newline character ('\n') when you use puts(). Although puts() is convenient to use, printf() has more formatting options, as shown in program P10A.

注意，使用函数puts( )时，无须加换行符'\n' 。虽然函数puts( )使用方便，但是从程序P10A可以看出，函数printf( )比函数puts( )有更多的输出格式可供选择。

To read a string from the keyboard, scanf() is used with a %s in the format string. The next program reads in a string using scanf() and displays it using puts().

为了从键盘读取一个字符串，需要在函数scanf( )的格式字符串中使用%s格式转换符。

Program Example P10B

```
1  /* Program Example P10B
2     This program inputs a name and displays it using puts(). */
3  #include <stdio.h>
4  int main()
5  {
6    char in_name[21] ; /* in_name will hold up to 20 characters + 1
7                          for the terminating null character '/0'. */
8
9    printf( "Type your name:" ) ;
10   scanf( "%s", in_name ) ; /* Note that the & is not necessary. */
11
12   printf( "Hello, " ) ;
13   puts( in_name ) ;
14   return 0 ;
15 }
```

Note that the scanf() function on line 10 does not contain &in_name. This is because in_name is an array, and the name of an array is a pointer to the first element of the array, i.e. in_name is equivalent to &in_name[0].

注意，程序第10行的scanf( )函数并未在in_name前添加取地址运算符&，这是因为in_name本身是一个数组，而数组名是指向数组第一个元素的指针，即in_name等价于&in_name[0]。

The following is a sample run of this program:

```
Type your name:John
Hello, John
```

Another sample run of this program is:

```
Type your name:Zhang Hui
Hello, Zhang
```

Why was Hui not displayed?

为什么名字Hui没有显示出来呢？

The %s in the format string instructs scanf() to read the characters up to, but not including, a whitespace character and place them in the array in_name with a '\0' after the last character. The remaining characters are left in the keyboard buffer and are not read by this program.

这是因为格式字符串中的%s格式转换符指示scanf( )函数在读取字符串到字符数组in_name中时，遇到空白字符（不包括空白字符）就结束了输入，然后将字符数组in_name 的最后一个字符置为'\0'，空白字符后剩余的字符仍然保留在输入缓冲区中，未被程序读入。

Line 6 of program P10B specifies any array size of twenty-one (twenty characters for the name plus one for the terminating character '\0'). If you type in a name of more than twenty characters, the excess characters will overwrite other areas of memory, and the program will probably malfunction. To solve this problem, specify a width in the format string of scanf() as in:

程序P10B的第6行语句指定数组的大小为21，其中前20个元素用于存放姓名，剩下的1个元素用于存放字符串结束符'\0'。如果输入的字符数大于20，那么多余的字符将覆盖数组之后的内存空间，程序很可能就会失效。为了解决这个问题，我们可以在scanf( )函数的格式字符串中指定域宽。

```
scanf( "%20s", in_name ) ;
```

The %20s instructs scanf() to read a string of characters, starting at the first non-whitespace character and ending after twenty characters are read or a whitespace character is encountered, whichever comes first.

格式转换符%20s指定函数scanf( )从第1个非空白字符开始读入字符串，读入20个字符后，结束输入，如果在读满20个字符之前遇到空白字符，那么也结束输入。

10.6 The gets() function[gets( ) 函数]

Just as you can output a string using printf() or puts(), you can input a string using either scanf() or gets(). The gets() function will read all characters, including whitespace characters, up to the newline character. A newline character is generated by pressing the Enter key. Compare this with the behaviour of scanf(). The difference between gets() and scanf() is that scanf() with the format specifier %s reads a word from the keyboard, whereas gets() will read an entire line.

就像可以使用函数printf( )或puts( )输出字符串一样，同样也可以使用函数scanf( )或gets( )来输入一个字符串。函数gets( )将读取换行符之前的包含空白字符在内的所有字符。通过按回车键产生一个换行符。通过比较函数gets( )与函数scanf( )，我们发现二者的区别在于，函数scanf( )用%s格式符可以从键盘输入一个单词（不包括空格），而函数gets( )则可以从键盘输入一整行字符（可以包含空格）。

The next program is a rewrite of P10B using gets() rather than scanf().

Program Example P10C

```
1   /* Program Example P10C
2      To read a string from the keyboard using gets(). */
3   #include <stdio.h>
4   int main()
5   {
6     char in_name[21] ; /* in_name will hold up to 20 characters + 1
7                           for the terminating null character '/0'. */
8
9     printf( "Type your name:" ) ;
10    gets( in_name ) ;
```

```
11
12   printf( "Hello, " ) ;
13   puts( in_name ) ;
14   return 0 ;
15 }
```

A sample run of this program is:

```
Type your name:Zhang Hui
Hello, Zhang Hui
```

Note that `gets()` reads the entire line containing the two names. In program P10B, `scanf()` only read one of the names.

注意，这里函数gets( )读入了用户输入的一整行字符（包括姓和名），而在程序P10B中，函数scanf( )只读取了用户输入的姓名中的名。

10.7 Accessing individual characters of a string（访问字符串中的单个字符）

As a string is an array of characters, you can access each character of the string by using the subscripts of the array. The next program displays each character of the string `"hello"`.

由于字符串本身就是一个字符数组，因此可以使用数组下标来访问字符串中的每个字符。下面的程序使用下标法输出字符串“hello”中的每一个字符。

Program Example P10D

```
1  /* Program Example P10D
2     To demonstrate the use of subscripts in accessing each
3     individual character of a string.                           */
4  #include <stdio.h>
5  int main()
6  {
7    char greetings[6] = "Hello" ;
8    int i ;
9
10   /* Display each of the characters of greetings on a new line. */
11   for ( i = 0 ; i < 5 ; i++ )
12   {
13     printf( "%c\n", greetings[i] ) ;
14   }
15  return 0 ;
16 }
```

This program will display each letter of `greetings` on a separate line:

```
H
e
l
l
o
```

Instead of `greetings[i]` in line 13, the equivalent pointer expression `*(greetings+ i)` could also be used.

10.8 Assigning a string to a pointer （用字符串为字符指针赋值）

In C a string literal is an array of characters, and the name of an array is a pointer to its first element. Therefore, C treats a string literal as a pointer to the first character in the string.

在C语言中，字符串本身就是一个字符数组，而数组名又是指向它的第一个元素的指针，因此可以将字符串看成是指向字符串第一个字符的指针。

As a string literal is a pointer, it is possible to assign a string to a `char` pointer. For example:

既然字符串可看成是一个指针，那么就可以将其赋值给一个字符型指针变量。

```
char *p ;
p = "some text" ;
```

or

```
char *p = "some text" ;
```

Here `p` is defined as a `char` pointer and assigned the address of the character string `"some text"`. The memory locations will look something like the following sketch as Figure 10.3.

这里，p定义为一个字符型指针变量，并且被赋值为字符串"some text" 的首地址。

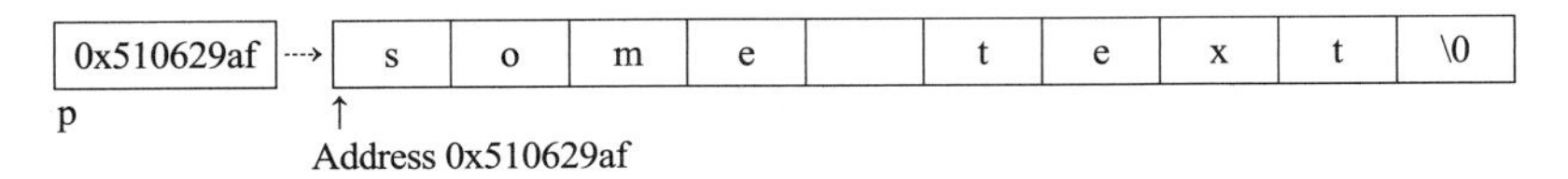

Figure 10.3 a pointer to a string

The string `"some text"` is held somewhere in memory, for example at the hexadecimal address 0x510629af. The pointer variable `p` holds the address (0x510629af) of the string.

字符串"some text"存储在内存中的某个区域内，例如首地址为0x510629af开始的一段连续内存空间，指针变量p中保存了该字符串的首地址0x510629af。

The next program displays each character of the string `"some text"` on a separate line using a pointer.

Program Example P10E

```
1  /* Program Example P10E
2     This program displays each character of the string "some text"
3     on a new line using a pointer variable.                        */
4  #include <stdio.h>
5  int main()
6  {
7    char *p = "some text" ; /*p points to the first character 's' */
8
9    /* The next while loop is performed until p points to the null
10      character '\0' at the end of the string. */
11   while ( *p != '\0' )
12   {
13     printf( "%c\n", *p ) ; /* Display each character. */
14     p++ ; /* p now points to the next character.      */
15   }
16   return 0 ;
17 }
```

The definition

```
char *p = "some text" ;
```

is subtly different from the definition:

```
char s[] = "more text" ;
```

Although p and s are pointers, the difference between them is that p is a variable pointer and s is a constant pointer. Unlike p, s does not occupy a memory location, as shown by the following memory sketch as Figure 10.4.

虽然p与s同为指针，但不同的是，p为指针变量，而s为常量指针，p占用一定字节数的内存空间，而s本身并不占用内存空间。

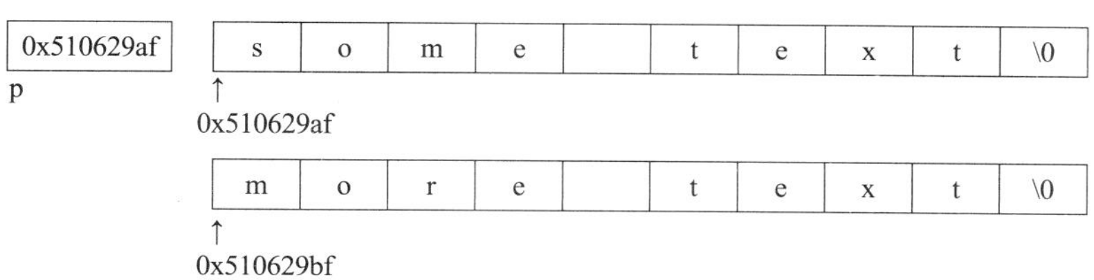

Figure 10.4 variable and constant pointers

You can assign s to p, as in

```
p = s ;
```

but not p to s, as in

```
s = p ; /* Illegal: s is a constant. */
```

10.9 String functions（字符串处理函数）

To use any of the standard library string functions you must include the preprocessor directive

```
#include <string.h>
```

at the beginning of the program.

使用C标准函数库中的字符串处理函数时，需在程序的开头用#include编译预处理指令包含string.h头文件。

10.9.1 Finding the length of a string

The standard library function **strlen( )** returns the number of characters in a string, excluding the null character '\0'.
Example:

标准库函数strlen( )的功能是：返回一个字符串中除空字符'\0'以外的所有字符的个数。

```
char name1[]   = "Sharon" ;
char name2[10] = "Mark" ;
char *name3    = "Xiaoling" ;
int len ;
len = strlen( name1 ) ;
printf( "%d  %d  %d  %d  %d",
        strlen( name1 ), strlen( name2 ),
        strlen( name3 ), strlen( "Rob" ), len ) ;
```

This will display: 6 4 8 3 6

The general format of the strlen() function is:

```
len = strlen( str )
```

where the argument `str` is a null-terminated string and `len` is an integer. (A null-terminated string is a string of characters ending with the null character `'\0'`.)

其中，参数str是一个包含字符串结束符的字符串（以空字符'\0'为字符串结束符），len是一个整型变量。

10.9.2 Copying a string

The string copy function, **strcpy(str1, str2)**, copies the contents of a string `str2` to `str1`.

字符串复制函数strcpy(str1, str2)的功能是：将字符串str2的内容复制到str1中。

Example:

```
char name1[] = "Sharon" ;
char name2[9] = "Tiantian" ;
/* Copy the contents of name1 to name2. */
strcpy( name2, name1 ) ;
/* Restore the original name. */
strcpy( name2, "Tiantian" ) ;
```

The general format of `strcpy()` is:

```
strcpy( destination, source ) ;
```

where the source string is copied to the destination string. The source string must be null-terminated. The `strcpy()` function assumes that the destination string is big enough to hold the string being copied to it. No checking is performed, so beware!

该函数将源字符串source的内容复制到目的字符串destination中。源字符串必须以空字符作为结束符，函数strcpy( )假设目的字符串所占用的内存空间足够大，能容纳得下待复制的源字符串的内容，所以它不进行任何相关的检测。在使用该函数时要特别注意这一点。

10.9.3 String concatenation

The standard library function **strcat(str1, str2)** concatenates a string `str2` to the end of the string `str1`. Both `str1` and `str2` must be null-terminated. The null character `'\0'` at the end of `str1` is removed before `str2` is concatenatedto `str1`. The null character `'\0'` is then added to the end of the new string in `str1`. Therefore, enough memory must be allocated to the string `str1` for it to hold the result of the concatenation.

标准库函数strcat(str1, str2)的功能是：将字符串str2的内容连接到字符串str1的尾部。字符串str1和字符串str2都必须以空字符为结束符。在将str2连接到str1后面之前，str1末尾的空字符'\0'被移除，在连接后的新字符串str1的后面添加空字符'\0'。因此，必须给字符串str1分配足够大的内存空间，以便能容纳得下连接后的字符串。

Example:

```
char str1[17] = "first and " ;
char str2[] = "second" ;
strcat( str1, str2 ) ; /* str1 now contains the string
                          "first and second". str2 is
                          unchanged.                 */
```

10.9.4 Comparing strings

The standard library function **strcmp(str1, str2)** compares two null-terminated strings `str1` and `str2`. This function returns a negative value if the string in `str1` is less than the string in `str2`, 0 if the string in `str1` is equal to the string in `str2`, and a positive value if the string in `str1` is greater than the string in `str2`.

Example:

标准库函数strcmp(str1, str2)的功能是：比较两个以空字符作为结束符的字符串str1和str2的大小，如果字符串str1小于字符串str2，则函数返回一个负值，如果字符串str1与字符串str2相等，则函数返回0

```
char password[7] = "secret" ;
char user[81] ;
gets( user ) ;
if ( strcmp( password, user ) == 0 )
  printf( "Correct password. Welcome to the system ...\n") ;
else
  printf( "Invalid password.\n" ) ;
```

In this section of code a user types in a password. The password entered is stored in `user` and is compared with an internal password held in `password`. The function `strcmp()` will return 0 if there is an exact match and a welcome message is displayed; otherwise an error message is displayed.

值，如果字符串str1大于字符串str2，则函数返回一个正值。这段代码要求用户输入一个口令，然后将其存入字符数组user中，并将其与保存在字符数组password中的系统内部口令进行比较，如果它们恰好相等，则函数strcmp( )返回0值，同时程序显示欢迎信息，否则显示出错信息。

10.9.5 Other string functions

The following are just some of the more commonly used string functions available in the standard library.

strchr(str, ch)
Finds the first occurrence of a character `ch` in a string `str`. This function returns a pointer to the first occurrence of `ch`. If `ch` is not in `str`, the `NULL` value is returned.

查找字符ch在字符串str中首次出现的位置，函数返回指向字符串str中首次出现的字符ch的指针，如果字符串str中不存在字符ch，则返回空指针NULL。

strncat(str1, str2, n)
Appends the first `n` characters of the string `str2` to the string `str1`.

将字符串str2的前n个字符添加到字符串str1的末尾。

strncmp(str1, str2, n)
Identical to `strcmp( str1, str2 )`, except that at most `n` characters are compared.

函数strncmp(str1, str2, n)的功能与函数strcmp(str1, str2)类似，区别仅在于，strncmp(str1, str2, n)最多比较n个字符。

strncpy(str1, str2, n)
Copies `n` characters of the string `str2` into the string `str1`.

将字符串str2的前n个字符复制到字符串str1中。

strrchr(str, ch)
Same as `strchr( str, ch )`, except that the last occurrence of the character `ch` is found.

函数strrchr(str, ch)的功能与函数strchr(str, ch)类似，区别仅在于，strrchr(str, ch)是查找字符ch在字符串str中最后出现的位置。

strstr(str1, str2)
Finds the first occurrence of the string `str2` in the string `str1`. This function returns a pointer to the found string in `str1`. If `str2` is not in `str1`, the `NULL` pointer is returned.

查找字符串str2在字符串str1中首次出现的位置，函数返回指向在str1中首次找到的字符串str2的指针，如果字符串str1中不包含字符串str2，则函数返回空指针NULL。

10.10 Converting numeric strings to numbers（数值字符串向数值的转换）

Each character of the string "123" is stored in one byte of memory in the ASCII representation, as shown in Figure 10.5 below.

字符串“123”中的每个字符都以ASCII码的形式存储在

```
Character:
ASCII value in decimal:
ASCII value in binary:
```

'1'	'2'	'3'	'\0'
49	50	51	0
00110001	00110010	00110011	00000000

Figure 10.5　the string 123

内存中，占用一个字节的内存空间。

This is very different from the way in which an integer value of 123 is stored. Integer values are held in binary, not ASCII format. An integer value of 123 is represented in binary as shown in Figure 10.6 below.

00000000	01111011

Figure 10.6　binary value of the integer 123

字符串的存储方式与整型数123的存储方式截然不同，整型数是以二进制的形式存储在内存中的，而非ASCII码的形式。

The standard library functions **atoi( )** and **atof( )** convert a numeric ASCII string to its binary equivalent. To use any of these functions, you must include the preprocessor directive:

```
#include <stdlib.h>
```

at the beginning of your program.

Example:

```
char str[] = "123" ;
int int_number ;
double double_number ;

int_number = atoi( str ) ;     /* String to an integer.     */
double_number = atof( str ) ; /* String to a double float. */
```

These functions will ignore any leading whitespace characters and stop converting when a character that cannot be part of the number is reached. For example, `atoi()` will stop when it reaches a decimal point, but `atof()` will accept a decimal point, because it can be part of a floating-point number.

To check for invalid characters in the string, `strtof()` can be used to convert a string to a `float` and `strtod()` can be used to convert a string to a `double`.

Example:

C语言的标准库函数atoi( )和atof( )的功能是，将数值型的以ASCII码形式存储的字符串转换为它的二进制等价形式。使用这些函数时，需在程序的开头用#include编译预处理指令包含头文件stdlib.h。

上述函数将忽略位于字符串前部的空白字符，当其读到某个字符并认为它不可能成为数值的组成部分时，转换结束。例如，由于函数atoi( )要将字符串转换为整型数值，因此当其读到小数点时，函数停止转换。但是因为小数点能够成为浮点数的组成部分，因此函数atof( )读到小数点时，能够将其转换。

```
char str[] = "123.5";
float float_number;
double double_number;
char *error_ptr;

float_number=strtof( str, &error_ptr ) ; /* String to a float.  */
if ( *error_ptr != '\0' )
  printf( "Error:%s is invalid, error is at the start of %s\n",
          str, error_ptr);
```

```
 double_number=strtod( str, &error_ptr) ; /* String to a double. */
if ( *error_ptr != '\0' )
  printf( "Error:%s is invalid, error is at the start of %s\n",
          str, error_ptr );
```

Both strtof() and strtod() take a second argument which will point to the first invalid character in the string, if one exists. If all the characters in the string are valid, then the second argument will point to the null character '\0'. This allows for error checking as shown with the if statement in the above code.

调用函数strtof()和strtod()时，如果字符串中存在无效字符，则函数的第二个实参指向字符串中的第一个无效字符（即用函数的第二个实参指出字符串中第一个无效字符的存储地址）；如果字符串中的全部字符都是有效字符，则第二个实参指向空字符'\0'，这样就可以用上面代码中的if语句通过判断第二个实参指向的字符是否为'\0'进行错误检查。

The standard library function strtol() converts a string to a long integer and also does error checking.

标准库函数strtol()的功能是将字符串转换为长整型并且进行错误检查。

Example:

```
  char str[] = "123456" ;
  long long_number ;
  char *error_ptr;

   long_number = strtol( str, &error_ptr, 10 ) ;
  if ( *error_ptr != '\0' )
    printf( "Error:%s is invalid, error is at the start of %s\n",
            str, error_ptr ) ;
```

As with strtof() and strtod(), strtol() uses a second argument to point to the first invalid character in the string, if one exists. The third argument used by strtol() is the base of the number contained in the string.

和函数strtof()、strtod()一样，函数strtol()使用第二个实参指向字符串中的第一个无效字符（假设存在）。第三个实参代表要转换的进制的基数，即表示要将字符串转换为几进制数。

Note: The standard library does not contain a function strtoi() to convert a string to an int. However, an int variable may be used in place of a long variable to store the value returned from strtol().

注意：标准库函数中不包含将字符串转换为int型的函数strtoi()。但是可换用long型变量代替int型变量来保存函数 strtol()的返回值。

10.11 Arrays of strings（字符串数组）

As a character string is an array of chars and the name of an array is a pointer to its first element, it follows that an array of strings is an array of pointers to chars.

如前所述，字符串本身就是一个字符型的数组，并且数组名可看成是指向数组第一个元素的指针，因此，字符串数组就是元素为字符型指针的数组。

The next program stores the months of the year in an array and displays them.

下面的程序将一年中的每个月份存储到一个字符串数组中，然后将其显示到屏幕上。

Program Example P10F

```
1  /* Program Example P10F
2     To demonstrate an array of strings. */
3  #include <stdio.h>
4  int main()
5  {
6    /* Define an array of strings. */
7    char *months[12] = { "January", "February", "March",
```

```
8                          "April", "May", "June", "July",
9                          "August", "September", "October",
10                         "November", "December" } ;
11    int i ;
12
13    /* Display the months of the year. */
14    printf( "The months of the year are:\n" ) ;
15    for ( i = 0 ; i < 12 ; i++ )
16      printf( "%s\n", months[i] ) ;
17    return 0 ;
18 }
```

Lines 7 to 10 of program P10F define an array of pointers to `chars`. Each element of this array contains the address of the first letter of a string. The strings themselves are stored elsewhere in memory. For instance, the first element, `months[0]`, contains the address of the string `"January"`, the second element, `months[1]`, contains the address of the string `"February"`, and so forth.

The loop in lines 15 to 16 displays each of the strings pointed to by each element of the array `months`. The output from this program will be as follows:

程序P10F的第7~10行定义了一个字符指针数组，数组的每个元素都保存了一个字符串首字母的地址值，而字符串存储在内存中与数组不同的位置。例如，数组的第1个元素months[0]中存放的是字符串"January"的首地址，数组的第2个元素months[1]中存放的是字符串"February"的首地址，以此类推。

```
January
February
March
April
May
June
July
August
September
October
November
December
```

The next program further demonstrates arrays of strings by reading five words of different lengths from the keyboard and displaying the longest word. Run this program and then read it carefully to understand how it works.

下面的程序进一步演示了字符串数组的用法，程序先要求从键盘输入5个长度不同的单词，然后输出长度最大的单词。

Program Example P10G

```
1  /* Program Example P10G
2     This program inputs five words of different
3     lengths and display the longest.          */
4  #include <stdio.h>
5  #include <string.h>
6  int main()
7  {
8    char word[5][81] ; /* An array to hold the five words. */
```

```
9     int index, index_of_longest ;
10    unsigned int longest_len = 0 ;
11
12    /* Input the words from the keyboard. */
13    for ( index = 0 ; index < 5 ; index++ )
14    {
15      printf( "\nEnter word%d: ", index+1 ) ;
16      scanf( "%s", word[index] ) ;
17    }
18
19    /* Now find the longest word. */
20    for ( index = 0 ; index < 5 ; index++ )
21    {
22      if ( strlen( word[index] ) > longest_len )
23      {
24        longest_len = strlen ( word[index] ) ;
25        index_of_longest = index ;
26      }
27    }
28
29    printf( "\nThe longest word is %s\n", word[index_of_longest]) ;
30    return 0 ;
31 }
```

A sample run of this program is:

```
Enter word1: supercalifragilisticexpialidocious
Enter word2: Haohaoxuexitiantianxiangshang
Enter word3: pneumonoultramicroscopicsilicovolcanoconiosis
Enter word4: Qianlizhixingshiyuzuxia
Enter word5: honorificabilitudinitatibus
The longest word is pneumonoultramicroscopicsilicovolcanoconiosis
```

Programming pitfalls

1. Double quotation marks (") are used for strings, single quotation marks (') are used for characters.
   For example, "a" is a string of two characters but 'a' is only one character.
2. When you are using scanf() to read a string there is no need to use the address operator &.
3. When you allocate space for a string you must allow for the terminating null character '\0'. For example, to store a string of twenty characters you must define an array with twenty-one elements.
4. Be careful when you are using pointers or subscripts that you stay within the bounds of the string (array).
5. Do not try to copy a string to a pointer that is not initialised to a memory location. For example:

```
char *p ;
char s[] = "abcdef" ;
strcpy( p, s ) ; /* Invalid: where is p pointing to? */
```

6. The name of an array is a constant pointer. It cannot be changed.
   For example:

```
char a[10] ;
a++ ;    /* Invalid: a is the name of an array. */
```

7. When you are using strcpy(str1, str2), the string str2 is copied to the string str1, and not the other way around.
8. The null character is '\0', not '/0'.
9. To compare two strings you need to use strcmp(). For example,

```
char str1[] = "ABCD" ;
char str2[] = "ABCD" ;
```

   The statement

```
if ( str1 == str2 ) /* Comparing pointers. */
```

   will only compare the constant pointers str1 and str2, which will not contain the same address. The result of the if statement, therefore, will always be false.
   To compare the actual strings, strcmp() is used.

```
if ( strcmp( str1, str2 ) == 0 ) /* Comparing strings. */
```

10. When displaying most strings you may omit the %s format specification in printf(). For example, if str is a string, then

```
printf( str ) ;          /* Works most times. */
```

   and

1. 字符串是用双引号括起来的，而单个字符是用单引号括起来的。例如，"a"是有两个字符的字符串，而'a'则是单个字符。
2. 当使用scanf( )函数读入字符串时，无须添加取地址运算符&。
3. 为字符串申请内存空间时，一定要为字符串结束符'\0'多申请一个字节的空间。例如，为了存储包含20个字符的字符串，必须定义一个长度为21的字符数组。
4. 使用指针或下标访问数组元素时，注意指针值或下标值不要超出字符串（数组）的边界。
5. 不要将字符串赋值给一个未被初始化的指针变量。
6. 数组名是一个常量指针，它的值是不可以改变的。
7. 注意函数strcpy(str1, str2)是将字符串str2的值复制到字符串str1中，而不是反过来将str1的值复制到str2中。
8. 字符串中的空结束符是'\0'，而不是'/0'。
9. 比较两个字符串大小时，应该使用函数strcmp( )。
10. 在多数情况下，使用函数printf( )输出字符串时，可以省略格式转换符%s。如果字符串str中含有%字符，那么使用第一种方式

```
printf( "%s", str ) ;    /* Better to use this. */
```

are nearly, but not actually, equivalent. Strange things may happen with the first statement if `str` contains `%` characters. This is because `printf()` interprets the character after a `%` as a format specification. You can try this out by defining `str` as a string such as the following:

```
char str[] = "This string contains a %s and a %d" ;
```

11. `scanf()` cannot detect the end of the array that you have allocated for a string and may store characters past the end of the array if too many characters are entered at the keyboard. `scanf()` can be made safer by using `%ns` instead of `%s` (`n` is the number of characters in the array and is a constant).

将输出奇怪的结果，这是因为printf( )函数中的%与其后面的字符连在一起，将被解释为格式转换符。

11. scanf( )函数不对为字符数组分配的内存空间进行边界检查。如果用户从键盘输入的字符数过多，那么scanf( )函数可能会将字符存储到超出数组边界以外的内存单元中。将格式转换符%s替换为%ns（n为常量，其值等于数组中可容纳的字符个数），会使scanf( )函数在读取字符串时更安全。

Quick syntax reference

	Syntax	Examples

Defining a character string.	`char string[N] ;`
	N is the number of characters in the string including the null character "\0". *N* is a positive constant integer, not a variable.
		`char str[11] ;`
		`/*`
		`str can hold ten characters plus the null character '\0'.`
		`*/`

String input from the keyboard.	`scanf("%s",string);/* no & */` `gets(string) ;`	`scanf("%s", str) ;` `gets(str) ;`
Display a string.	`printf("%s", string) ;` `puts(string);`	`printf(%s, str) ;` `puts(str) ;`
String length.	`size=strlen(string);`	`len=strlen(str) ;`
Comparing strings.	`strcmp(string1,string2);`	`if (strcmp(s1, s2)==0)` `  puts("Identical") ;`
Copying a string.	`strcpy(destination, source);`	`strcpy(s1, s2) ;` `/* Copy s2 to s1. */`

Exercises

1. What do the following `printf()` statements display?

   (a) `printf( "%5s", "abcd" ) ;`

   (b) `printf( "%5s", "abcdef" ) ;`

   (c) `printf( "%-5s", "abc" ) ;`

   (d) `printf( "%5.2s", "abcde" ) ;`

   (e) `printf( "%-5.2s", "abcde" ) ;`

2. Given the following array definition,

```
char name[] = { 'R','o','b','e','r','t' } ;
```

what is wrong with the following statements?

(a) `puts( name ) ;`

(b) `scanf( "%s", &name ) ;`

(c) `strcpy( name, "Philip" ) ;`

(d) `*( name + 5 ) = "a" ;`

(e) `name = "Peiqi" ;`

3. Given the following:

```
char *text = "some text" ;
```

what is the output from each of the following?

(a) `printf( "%s\n", text ) ;`

(b) `printf( "%c\n", *text ) ;`

(c) `printf( "%c\n", *(text+1) ) ;`

(d) `printf( "%s", text+1 ) ;`

(e) `printf( text ) ;`

(f) `*( text + 4 ) = '\0' ; printf("\n%s\n", text) ;`

4. Given the following:

```
char str1[21] ;
char str2[10] = "                " ;
```

what is in `str1` after each of the following?

(a) `strcpy( str1, "a string" ) ;`

(b) `strcat( str1, " of text." ) ;`

(c) `strncpy( str1, str2, 2 ) ;`

(d) `strncpy( str1+2, str2+1, 2 ) ;`

5. What is the output from the following two program segments?

(a)
```
char *p = "abcd" ;
while ( *p )
    putchar( *p++ ) ;
```

(b)
```
char *text = "abcd" ;
char *p = text ;
p += strlen( p ) - 1 ;
while ( text <= p )
  puts( p-- ) ;
```

6. Write a program to read in your name and display it with a space between each letter. For example, Lixiaolong gets displayed as L i x i a o l o n g.

7. Write a program to read in each word of a sentence and calculate the average length of the words in that sentence. You can assume that a sentence ends with a full stop and each word is separated from the next by a space. You can also assume that no word is more than twenty-five characters long.

8. Modify exercise 7 to display the number of words in the sentence with lengths of

(a) 1

(b) 2 to 5

(c) 6 to 10

(d) 11 to 20

(e) 21 and above.

9. Write a program to ask a user for their name. The user's name is then compared with a list of names held in an array in memory. If the user's name is in this list, display a suitable greeting; otherwise display the message "Name not found".

10. Write a program to input a four-character string. If every character in the string is a digit (0 to 9), then convert the string to an integer, add 1 to it, and display the result. If any one of the characters in the string is not a digit, display an error message.

11. The following is a list of countries and their capital cities.

Australia	Canberra
Belgium	Brussels
China	Beijing
Denmark	Copenhagen
England	London
France	Paris
Greece	Athens
Ireland	Dublin
Scotland	Edinburgh
Wales	Cardiff

Write a program to input a country and display the capital city of that country.

Chapter Eleven
Functions
第 11 章　函　　数

11.1 Introduction（引言）

Functions are the building-blocks of a C program. A function is a named block of statements carrying out a specific task, such as displaying headings at the top of every page of a report, reading a file, or performing a series of calculations. C has a variety of functions in the standard library, some of which you have already used, for example `printf()`, `scanf()`, and `strlen()`. The real power of C (or any other programming language) lies in the facility to write your own functions to add to those already at hand in the standard library.

As programs become more complex, they become difficult, even impossible, to write without the use of functions. To solve a complex problem the programmer needs to break down a problem into smaller, more manageable tasks. This is called *decomposition*, and the use of functions allows the programmer to do this.

By way of introducing functions, the next program displays a string of text with a line of stars (asterisks) above and below it.

函数是构建C程序的基本模块，是用于执行特定任务的命名的语句块。例如，在报告每页的顶端显示标题、读取文件、执行一系列的计算任务等。C 语言在标准函数库中提供了多种用于完成不同功能的函数，其中的某些函数已在前面的章节中使用过，例如printf( )、scanf( )、strlen( )等。

C语言或其他编程语言的真正魅力在于，它具有允许程序员自己编写函数并将其添加到现有的标准函数库中的能力。

随着程序复杂性的提高，不使用函数来编写程序是非常困难的，甚至是不可能的。为了解决一个复杂的问题，程序员需要将其划分为一系列更小、更易解决的子问题，这个过程称为**问题分解**。程序员通过使用函数就可以实现这一过程。

Program Example P11A

```
1  /* Program Example P11A
2     To demonstrate the use of functions by displaying
3     a string of text eclosed within two lines of stars.    */
4 #include <stdio.h>
5 void stars( void ) ;   /* The declaration of the function. */
6 int main()
7 {
8   char text[10] = "some text" ;
9
10  stars() ; /* Call the function to display the top line. */
11  printf( "%s\n", text ) ;
12  stars() ; /* Call the function again for the bottom line. */
13  return 0 ;
14 } /* This is the end of the main function. */
15
16 /* The function definition. */
```

```
17 void stars(void)
18 {
19   int counter ;
20
21   for ( counter = 0 ; counter < 9 ; counter++ )
22     putchar( '*' ) ;
23   putchar ( '\n' ) ;
24 } /* This is the end of the stars function. */
```

When this program is run you will get the following displayed on your screen:

```
*********
some text
*********
```

Like variables, functions must be declared before they are used. However, unlike variables, functions are normally declared before `main()`.

In this program, line 5 declares `stars` to be a function. The first **void** on line 5 declares the type of the function `stars()`. Functions with type `void` do not return a value to the calling program. Other functions do return a value: for example, the `strlen()` function returns an integer value, i.e. the length of a string. If you do not declare a type for a function, C assumes that it will return an integer value.

The `void` in parentheses on line 5 informs the compiler that the function `stars()` will not receive any data from the calling program. Many of the functions that you have used previously, require data in the parentheses. One such function is `strlen()`, which requires string data to be passed to it. However, the function `stars()` does not require any data to be passed to it.

In summary:

```
void stars( void ) ;
↑              ↑
```

returns nothing receives nothing

The function declaration in line 5 is also called the *prototype* of the function `stars()`.

Lines 10 and 12 call the function `stars()`, resulting in the top and bottom lines of stars being displayed.

Lines 17 to 24 define the function. The function definition begins with the lines

```
void stars()
{
```

```
/* and ends with the line */
}
```

与变量类似，调用函数前必须对其进行声明。但是不同于变量的是，函数通常在main( )函数之前进行声明。

这个程序的第5行将stars声明为一个函数。函数名前面的关键字void 声明函数的返回值类型为空类型，表明函数无须返回值给调用程序。另一些函数是有返回值的，例如strlen( )函数返回一个整型值，即字符串的长度。如果在函数声明中没有指定函数返回值的类型，那么C语言默认该函数的返回值类型为整型。第5行语句括号中的void的作用是，通知编译器函数不从调用程序中接收任何参数。之前使用的很多函数是需要在圆括号中给出数据的，例如strlen( )函数，需要给它传递字符串数据。而stars( )函数则不需要给它传递数据。

程序第5行的函数声明，也称为函数stars( )的**函数原型**。

This is the same way that `main()` is defined.

The variable `counter` defined in line 19 of program P11A is a *local* variable to the function `stars()`. Local variables are known only within the function in which they are defined. If a variable of the same name is defined in `main()` or in another function, C regards the two variables as different.

程序P11A第19行定义的变量counter是函数stars( )的**局部变量**，局部变量只能在定义它的函数范围内使用，如果在主函数或其他函数中定义了与该局部变量同名的变量，则C语言将其视为不同的变量。

In addition to the function `stars()`, program P11A also uses the `printf()` and `putchar()` functions. These functions are supplied with the compiler and their prototypes are in the file `stdio.h`, included in the program on line 4.

11.2 Function arguments（函数参数）

The function `stars()`, as written in P11A, displays nine stars every time it is called. It would be useful to have a function to display a variable number of stars, not just nine. This new function could be used for displaying lines of stars with different sizes.

如程序P11A所示，每次调用函数stars( )都将在屏幕上输出9个星号。编写一个能输出可变长度（不只是9个）的星号的函数会更有用。这个新函数可以用于输出不同长度的星号行。

The next program modifies the function `stars()` to receive a value passed to it by the calling program. The function then displays the number of stars specified in the passed value.

下面的程序对前面的stars( )函数进行了修改，通过接收调用程序传递给它的参数，可以按该参数值指定的数目在屏幕上输出星号。

Program Example P11B

```
1  /* Program Example P11B
2     Demonstration of calling a function with an argument. */
3  #include <stdio.h>
4  #include <string.h>
5  void stars( int ) ; /* The declaration of the function. */
6  int main()
7  {
8    char text[81] ;
9    int len ;
10
11   printf ("Type a string of text and press Enter: ");
12   scanf ( "%s", text) ;
13   len = strlen ( text ) ;
14   stars( len ) ; /* Call the function to display the top line. */
15   printf( "%s\n", text ) ;
16   stars( len ) ; /* Call the function again for the bottom line. */
17   return 0 ;
18 }
19
20 void stars( int num )
21 {
22   int counter ;
23
24   for ( counter = 0 ; counter < num ; counter++ )
25     putchar( '*' ) ;
26   putchar ( '\n' ) ;
27 }
```

A sample run of this program is:

```
Type a string of text and press Enter: TextString
**********
TextString
**********
```

Calls to the function `stars()` in lines 14 and 16 now have a number between the parentheses (and). This number is called the *argument* and is received by the *parameter* `num` in line 20.

Line 20 declares `num` to be an integer. The parameter or parameters of a function are declared in the parentheses as shown in line 20. Like local variables, the parameters of a function are known only within the function. Therefore, variables with the same name can be used in `main()` or in any other function without a conflict occurring.

Line 24 of the function `stars()` now uses the variable `num` to decide how many times * is displayed.

Since the function `stars()` is now receiving various integer values from the calling program `main()`, the declaration in line 5 now has `int` in the parentheses.

An even more useful function than `stars()` would be one where you could specify not only the number but also the character you want to display. The next program includes such a function called `display()`.

在程序第14行和第16行调用函数stars( )时，分别在圆括号中指定了一个数值，称为函数的实参，程序第20行的函数定义中的形参num接收该实参的值。

程序第20行将num声明为整型，函数的形参在程序第20行的圆括号中声明。与局部变量类似，函数的形参只能在声明它的函数中使用。因此，在主函数或其他函数中可以使用与其同名的变量，而不会发生冲突。

在函数stars( )的第24行语句中，使用变量num来确定要输出多少个星号。

因为函数stars( )要从调用它的主函数main( )中接收各种整型的数据，所以第5行的函数声明中应该在其圆括号内指定形参类型为int。

如果使用户不仅可以指定输出字符的个数，而且还可以指定要输出什么字符，那么这个函数将更加有用。下面的程序就包含了这样的一个函数disp_chars( )。

Program Example P11C

```
1  /* Program Example P11C
2     Demonstration of a function with two parameters. */
3  #include <stdio.h>
4  #include <string.h>
5  void display( int num, char ch) ; /* Function declaration. */
6  int main()
7  {
8    char text[81] ;
9    int len ;
10
11   printf ("Type a string of text and press Enter: ");
12   scanf ( "%s", text) ;
13   len = strlen ( text ) ;
14
15   display( len, '+' ) ; /* Display the top line of +s. */
16   printf( "%s\n", text ) ;
17   display( len, '-' ) ; /* Display the bottom line of -s. */
18   return 0 ;
19 }
20
21 /* Function : display
22    This function will display any number of any character.
```

```
23     Parameters: num is the number of times to display a character.
24                 ch is the character to display.                      */
25 void display( int num, char ch )
26 {
27   int counter ;
28
29   for ( counter = 0 ; counter < num ; counter++ )
30     putchar( ch ) ;
31   putchar ( '\n' ) ;
32 }
```

A sample run of this program is:

```
Type a string of text and press Enter: TextString
++++++++++
TextString
----------
```

The function `display()` uses two parameters: `num` (the number of times to display a character) and `ch` (the character to display). These two variables are declared in line 25. Values are passed to these parameters when the function is called in lines 15 and 17.

The function declaration in line 5 informs the compiler that the function `display()` will receive an `int` and a `char` from the calling program. The compiler checks that the type and number of arguments used in the function calls match the function prototype.

```
void disp_chars( int num, char ch ) ;
↑                 ↑         ↑
returns nothing   receives an integer and a character
```

Notice that this time variable names are included in the function declaration in line 5. The variable names are optional and can be any valid variable names that may or may not be defined elsewhere in the program. In practice, the variable names are often the same as the parameters used in the function. This means that the declaration (line 5) is often the same as the first line of the function definition (line 25), but the declaration ends with a semicolon.

函数display( )有两个形参：num（要显示的字符的个数）和ch（要显示的字符），程序的第25行对函数的两个形参进行了声明。程序在第15行和第17行调用函数时，将实参的值传递给相应的形参。

程序第5行的函数声明，通知编译器函数display( )从调用程序接收两个实参，一个是整型，另一个字符型。编译器会对在调用函数时传递的实参个数和类型与函数原型是否匹配进行检查。

注意，这一次在程序的第5行的函数声明中，给出了形参变量的名字。我们也可以在函数声明中将形参的名字省略不写，或者写成其他任何合法的变量名，但实际上，通常将该变量与函数定义时的形参取为相同的名字。也就是说，函数声明（例如程序的第5行）与函数定义（例如程序的第25行）的首行是相同的，只不过是函数声明的末尾多了一个分号。

11.3 Returning a value from a function（从函数返回一个值）

To demonstrate returning a value from a function, the next program uses a function that takes three integer values as parameters and returns the minimum of the three values.

Program Example P11D

```
1  /* Program Example P11D
2     To demonstrate the use of the return statement in a function. */
```

```
3 #include <stdio.h>
4 int minimum(int num1, int num2, int num3);/*Function declaration.*/
5 int main()
6 {
7   int val1, val2, val3, min_val ;
8
9   /* Read in three integer values from the keyboard. */
10  printf( "Please enter three integers: " ) ;
11  scanf( "%d %d %d", &val1, &val2, &val3 ) ;
12
13  /* Find the minimum of these three values. */
14  min_val = minimum( val1, val2, val3 ) ;
15  printf( "The minimum of %d, %d and %d is %d\n",
16           val1, val2, val3, min_val ) ;
17  return 0 ;
18 }
19
20 /* Function   : minimum
21    This function will return the minimum of three integer values.
22    Parameters: three integer values num1, num2 and num3.
23    Returns    : the minimum of num1, num2 and num2.          */
24 int minimum( int num1, int num2, int num3 )
25 {
26   int min = num1 ; /* Assume the first number is the minimum.  */
27   if ( num2 < min )
28     min = num2 ;   /* The second number is the new minimum.    */
29   if ( num3 < min )
30     min = num3 ;   /* The third number is the new minimum.     */
31   return min ;
32 }
```

Running this program will give the following result:

```
Please enter three integers: 3 2 1
The minimum of 3, 2 and 1 is 1
```

Because the function `minimum()` returns an integer value, you must tell the compiler this in two places in your program. The first place is in the function declaration in line 4, and the second place is in the function header in line 24.

必须在程序中的两个位置，告知编译器函数minimum( )返回一个整型数，一个是第4行的函数声明，另一个是第24行的函数定义的头部。

```
int minimum( int num1, int num2, int num3 )
 ↑             ↑__________↑__________↑
returns an integer   receives three integers
```

The `return` statement in line 31 does two things: it terminates the function and returns the value of the variable `min` (which is either `num1`, `num2` or `num3`) to the integer variable `min_val` in line 14.

The general format of the `return` statement is:

```
return expression ;
```

程序第31行的return语句做了两件事情：首先结束函数的运行，然后返回变量min的值（为num1、num2与num3三者之一），并通过第14行语句，将其赋值给整型变量min_val。

Although not required, many programmers enclose `expression` in parentheses:

尽管圆括号不是必须的，但很多程序员还是习惯使用圆括号将表达式expression括起来。

```
return ( expression) ;
```

Examples:

```
return 10.3 ;     /* Return a constant value. */
return ;     /* No return value, just exit the function.        */
return    variable ;  /* Return the value of a variable.        */
return ( variable + 1 ) ; /* Return the value of an expression. */
```

A function call can be used anywhere in a program where a variable can be used. For example, in the program P11D the variable `min_val` in line 16 may be replaced with the function call, as in:

函数调用可以出现在程序中变量可以出现的任何位置。例如，在程序P11D的第16行，变量min_val可以替换为函数调用。

```
printf( "Minimum of %d and %d is %d\n",
         val1, val2, minimum(val1, val2, val3) ) ;
```

11.4 Passing arguments by value（按值传参）

In the functions of the previous programs, a *copy of the values* of the arguments is passed to the parameters. This is known as *passing by value.*

Argument → Copy of *value* of argument → Parameter

As only a copy of the argument is sent to the function parameter, the value of the argument cannot be changed within the function.

前面程序中的函数都是将实参值的一个副本传递给函数的形参，这称为**按值传参**。由于函数调用时只是将实参值的一个副本传递给被调函数，因此实参的值在被调函数中是不能修改的。

Program Example P11E

```
1  /* Program example P11E
2     To demonstrate passing an argument by value to a function. */
3  #include <stdio.h>
4  void any_function( int a ) ;
5  int main()
6  {
7    int a = 1 ;
8
9    printf( "a is %d", a ) ;
10   /* Pass a copy of the value of a to the function. */
11   any_function( a ) ;
12   printf( "\na is still %d\n", a ) ;
13   return 0 ;
14 }
15
16 void any_function( int v )
17 {
18   printf( "\nThe value passed to the parameter is %d", v ) ;
19   v = 2 ; /* Change the value of the parameter. */
20   printf( "\nThe value of the parameter is now %d", v ) ;
21 }
```

When you run this program you will get the following:

```
a is 1
The value passed to the parameter is 1
The value of the parameter is now 2
a is still 1
```

This program changes the value of the parameter `v` in line 19 without having any effect on the argument `a`. Line 12 displays the value of `a`, showing it to be unchanged.

这个程序在第19行改变了形参v的值，但是形参v的值的改变对于实参a的值不会有任何影响。程序第12行输出a的值，说明了实参a的值并未改变。

11.5 Passing arguments by reference（按引用传参）

When you pass the address of one or more arguments to a function, it is known as *passing by reference*. Passing by reference allows you to change the value of the arguments by passing their addresses, rather than a copy of their values, to the parameters.

Argument → *Address* of argument → Parameter

在函数调用时向函数传递一个或多个实参的地址值，这种参数传递方式，称为**按引用传参**。按引用传参，使得我们可以在函数中通过传递实参的地址值来改变相应实参的值，而向函数形参传递实参值副本的传参方式不能在函数中改变实参的值。

Program Example P11F

```
1  /* Program Example P11F
2     To demonstrate passing an argument by reference to a function. */
3  #include <stdio.h>
4  void any_function( int *p ) ;
5  int main()
6  {
7    int a = 1 ;
8
9    printf( "a is %d", a ) ;
10   /* Pass the address of a to the function. */
11   any_function( &a ) ;
12   printf( "\na is now changed to %d\n", a ) ;
13   return 0 ;
14 }
15
16 void any_function( int *p )
17 {
18   printf( "\nThe value passed to the parameter is %d", *p ) ;
19   *p = 2 ; /* Change the value of the argument. */
20   printf( "\nThe value of the argument is now %d", *p ) ;
21 }
```

When you run this program you will get the following result:

```
a is 1
The value passed to the parameter is 1
The value of the argument is now 2
a is now changed to 2
```

Line 11 passes the address of the argument `a` to the function `any_function()` using the address operator `&`.

程序的第11行使用取地址运算符&将实参a的地址值传递给函数any_function( )。
由于函数any_function( )需要从程序第11行的函数调用中接收一个地址值，因此程序第16行将函数的形参声明为指针类型。

Line 16 declares the parameter to be a pointer, because it receives an address from the function call on line 11.

Line 19 uses the dereference operator `*` to change the value of `a`, and line 12 displays the value of `a`, showing that it has changed from 1 to 2.

程序的第19行通过使用解引用运算符*改变了程序实参变量a的值。程序第12行向屏幕输出a的值，由程序运行结果可以看出，该值由1变成了2。为了在函数内部改变实参的值，必须向函数传递实参的地址，也就是使用按引用传参的方式。

11.6 Changing arguments in a function （在函数中改变实参的值）

To change the value of the arguments from within a function, you must pass the addresses of the arguments to the function, i.e. pass by reference must be used.

The next program passes the addresses of two variables to a function `swap()`. This function swaps the variable values around, so that the first variable has the value of the second variable and the second variable has the value of the first variable.

下面的程序首先将两个变量的地址值传递给函数swap( )，在函数swap( )中实现两个变量值的互换，即用第一个变量保存第二个变量的值，而第二个变量保存第一个变量的值。

Program Example P11G

```
1  /* Program Example P11G
2     This program passes two arguments by reference to a function. */
3  #include <stdio.h>
4  void swap( float *ptr1, float *ptr2 ) ;
5  int main()
6  {
7    float num1, num2 ;
8
9    printf( "Please enter two numbers: " ) ;
10   scanf( "%f", &num1 ) ;
11   scanf( "%f", &num2 ) ;
12
13   if ( num1 > num2 )
14     swap( &num1, &num2 ) ; /* Swap the two values around. */
15   printf( "The numbers in order are %.1f  %.1f\n", num1, num2 ) ;
16   return 0 ;
17 }
18
19 /* Function   : swap
20    Purpose    : This function swaps two floating-point values.
21    Parameters: pointers to the variables to be swapped. */
22 void swap( float *ptr1, float *ptr2 )
23 {
24   float temp ;
25
26   temp = *ptr1 ;
27   *ptr1 = *ptr2 ;
28   *ptr2 = temp ;
29 }
```

A sample run of this program follows.

```
Please enter two numbers: 12.1 6.4
The numbers in order are 6.4  12.1
```

In line 4 the function `swap()` is declared as type `void`. This is because the function is not returning a value with a `return` statement.

由于函数swap( )没有使用return语句从函数返回值，因此在程序的第4行，我们将函数swap( )的返回值类型声明为void。

Line 14 passes the addresses of the two variables `num1` and `num2` to the function `swap()`. These addresses are received by the parameters `ptr1` and `ptr2`, declared as pointers to `float`s in line 22.

第14行将两个变量num1和num2的地址值传递给函数swap( )，这两个地址值被函数swap( )的两个形参ptr1和ptr2所接收。程序的第22行将这两个形参都声明为浮点型指针。

The statement

```
temp = *ptr1 ;
```

in line 26 stores the value of `num1` (= `*ptr1`) in the variable `temp` (`temp` is now 12.1).

The statement

```
*ptr1 = *ptr2 ;
```

in line 27 is equivalent to

```
num1 = num2 ;
```

because `*ptr1` is the same as `num1` and `*ptr2` is the same as `num2`. Therefore, `num1` gets the value 6.4.

Finally, the statement

```
*ptr2 = temp ;
```

in line 28 assigns the value of `temp` (i.e. 12.1) to `num2`, because `*ptr2` is the same as `num2`.

The overall result is that the values in `num1` and `num2` are swapped.

11.7 Passing a one-dimensional array to a function (向函数传递一维数组)

An entire array may be accessed within a function by using the name of the array as the argument when calling the function. This is done in the next program which includes a function `sum_array()` that computes the sum of the elements in an integer array.

在函数调用时，用数组名作为实参，可以实现在被调函数中访问整个数组元素的值。下面程序中的函数sum_array( )通过使用这种方法，实现了对一个整型数组中所有元素的求和。

Program Example P11H

```
1  /* Program Example P11H
2     To demonstrate passing a 1-dimensional array to a function. */
3  #include <stdio.h>
4  int sum_array( int array[], int num_of_elements ) ;
5  int main()
6  {
7    int values[10] = { 12, 4, 5, 3, 4, 0, 1, 8, 2, 3 } ;
8    int sum ;
9
10   sum = sum_array( values, 10 ) ;
```

```
11   printf( "The sum of the elements in the array is %d\n", sum ) ;
12   return 0 ;
13 }
14
15 /* Function    : sum_array
16    Purpose     : Sums the elements of a 1-dimensional integer array.
17    Parameters  : An array and the number of elements in the array.
18    Returns     : The sum of the elements in the array.            */
19 int sum_array( int array[], int num_of_elements )
20 {
21   int i, sum = 0 ;
22
23   for ( i = 0 ; i < num_of_elements ; i++ )
24      sum += array[i] ;
25    return sum ;
26 }
```

A run of this program will display:

```
The sum of the elements in the array is 42
```

Line 7 initialises the array `values`. Line 10 passes the name of the array (i.e. a pointer to the first element of the array) and the number of elements in the array to the function `sum_array()`.
Line 19 declares the parameters `array` and `num_of_elements`.
An alternative declaration for `array` is:

程序的第7行对数组values进行初始化，第10行将数组名（即指向数组第一个元素的指针）和数组元素的个数作为实参传递给函数sum_array( )。程序的第19行声明了函数的形参array和no_of_elements。

```
int *array
```

This is also valid because the name of an array is also a pointer to the first element of the array.
Line 24 uses subscript notation in accessing the elements of the array. Pointer notation can also be used, as in the following statement:

程序的第24行使用下标法访问数组元素，还可以使用指针法访问数组元素。

```
sum += *( array + i ) ;
```

Because the argument passed is a pointer to the first element in the array, the values of the elements in the array can be changed from within the function.
For example, placing the statements such as

由于传递给被调函数的实参是指向数组第一个元素的指针，因此数组元素的值可以在被调函数中被修改。

```
array[0] = 0    or   *array = 0 ;
array[1] = 0    or   *( array + 1 ) = 0 ;
```

in the function will assign `0` to the first two elements of the array.
The modifier `const` can be included in the declaration of the parameters in line 19:

在程序第19行的形参声明中，还可以在形参前面加上常量修饰符const。

```
int sum_array( const int array[], const int num_of_elements )
```

or

```
int sum_array( const int *array, const int num_of_elements )
```

These declarations inform the compiler that within the function sum_array, array and num_of_elements are read-only variables that cannot be modified. A compiler error will result if an attempt is made to change the value of a const variable.

The function declaration in line 4 must also change to:

```
int sum_array( const int *array, const int num_of_elements );
```

这样的声明相当于告知编译器，在函数sum_array( )内部，array和num_of_elements都是只读变量，它们的值是不能被修改的。如果程序试图修改const变量的值，那么将导致编译错误。

11.8 Passing a multi-dimensional array to a function（向函数传递多维数组）

When a multi-dimensional array is passed to a function, the parameter declaration must contain the size of each dimension of the array, except the first. Consider the following program, which sums the elements of a two-dimensional array.

向函数传递一个多维数组参数时，在函数的形参声明中，除数组第一维的长度可以不必指定外，其他维度的大小必须指定。

Program Example P11I

```
1  /* Program Example P11I
2     To demonstrate passing a 2-dimensional array to a function. */
3  #include <stdio.h>
4  int sum_array( int array[][2], int no_of_rows ) ;
5  int main()
6  {
7    int values[5][2] = { { 31, 14 },
8                         { 51, 11 },
9                         {  7, 10 },
10                        { 13, 41 },
11                        { 16, 18 } } ;
12   int sum ;
13
14   sum = sum_array( values, 5 ) ;
15   printf( "The sum of the elements in the array is %d\n", sum ) ;
16   return 0 ;
17 }
18
19 /* Function   : sum_array
20    Purpose    : Sums the elements of a 2-dimensional integer array.
21    Parameters: An array and the number of elements in the array.
22    Returns    : The sum of the elements in the array.            */
23 int sum_array( int array[][2], int no_of_rows )
24 {
25    int row, column ;
26    int sum = 0 ;
27
28    for ( row = 0 ; row < no_of_rows ; row++ )
29    {
30      for ( column = 0 ; column< 2 ; column++ )
31          sum += array[row][column] ;
```

```
32     }
33     return sum ;
34 }
```

This program displays the line:

```
The sum of the elements in the array is 212
```

The elements of a 2-dimensional array are stored row by row in contiguous memory locations. A sketch of the memory is shown in Figure 11.1.

二维数组中的元素在内存中是按行存储在一块连续的存储单元中的。

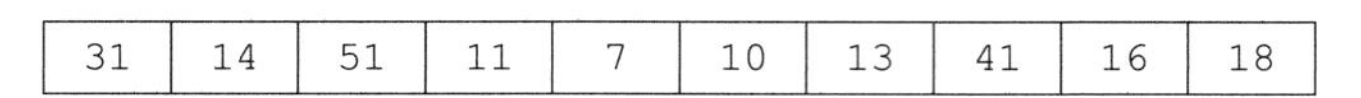

Figure 11.1 memory sketch for a 2-dimensional array

Any element `array[i][j]` has an offset i*2 + j from the starting memory location of `array[0][0]`. For example, `array[3][1]` (= 41) has an offset of 7 (= 3 * 2 + 1). In order to calculate the offset, the number of columns (= 2) is required; hence the need for the compiler to know the value of the second dimension in line 23.

Note: The parameter `array` could also be declared in line 23 using pointer notation as:

```
int ( *array )[2] ;
```

数组元素array[i][j]距离数组元素array[0][0]的地址偏移量为i*2+j。例如，array[3][1]（值为41）的偏移量为7（=3*2+1）。为了计算每个数组元素距离数组首地址的偏移量，编译器必须知道二维数组的列数（这里为2）。因此，在程序的第23行的形参声明中，必须指定数组的第二维的长度。

This declares `array` as a pointer to an array of two integer elements. (See Chapter Nine for a discussion of pointers and multi-dimensional arrays.) The parentheses are necessary in the above declaration. Without the parentheses the declaration

```
int *array[2]    ;  /* Error! */
```

incorrectly declares `array` to be an array of two pointers to integers.

11.9 Storage classes（变量的存储类型）

The variables used so far have been *automatic variables* with the storage type **auto**. This is just one of four types of storage available in C. The four types of storage are: **auto**, **static**, **register**, and **extern**. The default is `auto`.

到目前为止，我们用到的变量都是存储类型为auto的自动变量。在C语言中，变量的存储类型一共有4种，它们分别是自动类型auto，静态类型static，寄存器类型register和外部类型extern。系统默认的变量存储类型为auto。

11.9.1 auto

The variables defined inside a function are automatic by default. Every time a function is entered, storage for each `auto` variable is allocated. When the function is completed, the allocated storage is freed, and any values in the `auto` variables are lost. Such variables are known as *local variables* and are known only within the function in which they are defined. If you do not specify a storage class, `auto` is assumed.

系统默认为在函数内部定义的变量都是**自动变量**。当程序每次进入函数内部时，都要为自动变量分配内存空间，当函数执行结束时，释放为自动变量分配的内存空间，此时自动变量中存储的数据就会丢失。因此，自动变量是一种**局部变量**，它只能在定义它的函数内部使用。如果我们不指定变量的存储类型，那么系统将默认它为自动类型的变量。

For example:

```
void any_function()
{
  auto int var1 ;
  auto float var2[10] ;
  /* Function statements follow. */
  ...
}
```

The variables var1 and var2 have been defined with the storage class auto. As auto is the default, the keyword auto may be omitted. Automatic variables permit efficient use of storage, because the storage used by these variables is released for other purposes when a function exits.

变量var1和var2的存储类型定义为auto。由于系统默认的变量存储类型为auto，因此关键字auto可以省略。
当函数调用结束时，自动变量所占的内存被释放，可另为他用，因此使用自动变量有利于对内存空间的有效利用。

11.9.2 static

static variables, like auto variables, are local to the function in which they are defined. However, unlike auto variables, static variables are allocated storage only once and so retain their values even after the function is exited. The next program demonstrates the differences between auto and static variables.

与自动变量类似，静态变量也是局部变量，仅在定义它的函数中有效。但是，与自动变量不同的是，系统只对静态变量分配一次内存空间，所以当函数调用结束时，静态变量中存储的数据将仍然保留，不会丢失。

Program Example P11J

```
1  /* Program Example P11J
2     To demonstrate the difference between static and
3     auto variables.                                    */
4  #include <stdio.h>
5  void any_func( void ) ;
6  int main()
7  {
8    int i ;
9
10   /* Call the function any_func ten times. */
11   for ( i = 0 ; i < 10 ; i++ )
12   {
13     any_func() ;
14   }
15   return 0 ;
16 }
17
18 void any_func()
19 {
20   int auto_var = 0 ;
21   static int static_var = 0 ;
22
23   static_var++ ;  /* Increment the static variable. */
24   auto_var ++ ;   /* Increment the auto   variable. */
25   printf( "auto_var = %d  static_var = %d\n",auto_var,static_var );
26 }
```

The output from this program is:

```
auto_var = 1 static_var = 1
auto_var = 1 static_var = 2
auto_var = 1 static_var = 3
auto_var = 1 static_var = 4
auto_var = 1 static_var = 5
auto_var = 1 static_var = 6
auto_var = 1 static_var = 7
auto_var = 1 static_var = 8
auto_var = 1 static_var = 9
auto_var = 1 static_var = 10
```

The output from this program shows that the static variable `static_var` was initialised only once and retained its value between each function call. However, the `auto` variable `auto_var` was initialised to `0` every time the function was entered.

由程序的运行结果可知，静态变量static_var仅初始化了一次，并且在每次调用函数时，都能保持其原有的值，而自动变量auto_var在每次进入函数时都初始化为0值。

11.9.3 extern

External variables, like function arguments, allow you to transmit data between C functions. An external variable is defined outside the function definitions of a program and is available to every function in the program. For this reason, external variables are also known as *global variables*.

与函数参数类似，外部变量也可以在不同的C函数之间传递数据。外部变量在所有函数定义的外面定义，在所有函数中都能访问，正因如此，外部变量也称为**全局变量**。

Program Example P11K

```
1  /* Program Example P11K
2     Demonstration of external variables. */
3  #include <stdio.h>
4  void function1( void ) ;
5  void function2( void ) ;
6
7  int company_number = 1661234 ;
8  char company_name[] = "ABC & Company Ltd" ;
9
10 int main()
11 {
12   printf( "In main:\n" ) ;
13   printf( "company number = %d  company name = %s\n\n",
14            company_number, company_name ) ;
15
16   function1() ;
17   function2() ;
18   return 0 ;
19 }
20
21 void function1()
22 {
23   extern int company_number ;
24   extern char company[] ;
```

```
25   printf( "In function1:\n" ) ;
26   printf( "company number = %d  company name = %s\n\n",
27             company_number, company_name ) ;
28 }
29
30 void function2()
31 {
32   printf( "In function2:\n" ) ;
33   printf( "company number = %d  company name = %s\n\n",
34             company_number, company_name ) ;
35 }
```

The output from this program is:

```
In main:
company number = 1661234  company name = ABC & Company Ltd

In function1:
company number = 1661234  company name = ABC & Company Ltd

In function2:
company number = 1661234  company name = ABC & Company Ltd
```

Lines 7 and 8 define `company_number` and `company_name` as external global variables. These two variables are defined before `main()` and can be accessed in all functions in the program, including `main()`. The two variables are displayed in `main()` and also in the two functions `function1()` and `function2()`.

程序的第7行和第8行，将变量company_number和company_name定义为全局变量，这两个变量在主函数main( )的前面定义，可以在程序包括主函数main( )在内的所有函数中访问。

In lines 23 and 24, two variables are declared as external by using the keyword **extern**. This means that these variables are defined outside the function `function1()`, perhaps even in another program. If the external variables are defined within the same program file, the declaration of the variables in lines 23 and 24 is not necessary. For example, the variables `company_number` and `company_name` are not declared in the function `function2()`, but they are still accessible within the function and are used in line 34. Even though it is not always necessary, it is considered good practice to declare all external variables used within a function.

程序的第23行和第24行，使用关键字extern将这两个变量声明为外部变量，这意味着这两个变量是在函数function1( )之外，甚至可能是在其他程序中定义的。如果外部变量在同一个程序文件中被定义，那么就无须在程序第23行和第24行对外部变量进行声明。即使外部声明并非总是必须的，但在每个使用外部变量的函数中对外部变量进行声明，仍然被看成是一个好习惯。

Because global variables are known, and therefore can be modified, within every function, they can make a program difficult to debug and maintain. Global variables are not a substitute for function arguments. Strictly speaking, apart from its own local variables, a function should have access only to the parameters specified in the function prototype. However, it can be useful to have global variables for constants of a global

由于全局变量在程序的每个函数内都可以访问，因此其值在程序的每个函数内都可以修改，这给程序的调试和维护带来困难。因此，不宜用全局变量替代函数实参进行数据传递。严格地讲，除了函数内部定义的局部变量之外，函数只应访问在其原型中声明的形参。但是，可以使用全局变量保存程

nature, such as a constant data structure (see Chapter Twelve), the number of lines on a printer page, and so on.

Because the storage is always required for global variables, a program using global variables may require more memory than the same program using functions with parameters. Use global variables sparingly.

序中用到的具有全局性质的常量，例如常量数据结构（参见第12章）、每个打印页的行数等。由于在程序执行的过程中，全局变量一直占用着为其分配的内存，因此，对于同样的程序，使用全局变量传值比使用函数传参占用更多的内存，应少用、慎用全局变量。

11.9.4 register

The compiler attempts to place a variable with the storage class `register` in a CPU register rather than main storage. `register` variables are used to increase the speed of execution of a program.

Frequently in C programs, the index variable of a `for` loop is defined as a `register` variable. For example:

C编译器一般会试图将声明为寄存器存储类型的变量存储在CPU的寄存器中，而非主存中。因此，寄存器变量可以用来提高程序的执行速度。

在C程序中，常将for循环的索引变量定义为寄存器变量。

```
register int i ;
for ( i = 0 ; i < 1000 ; i++ )
```

Because there is a limited number of registers in a CPU, there is no guarantee that a variable defined as `register` will actually use a CPU register. If a `register` variable cannot be placed in a CPU register, the variable will have the default storage class `auto`.

Note: Many compilers try to optimise the speed of execution of a program by placing variables such as loop index variables in CPU registers. Most of the loop index variables in a program, therefore, will probably be register variables without you having to specify them as such.

但是，由于CPU中寄存器的数量是有限的，因此编译器不能保证被定义为寄存器存储类型的变量一定会存储到CPU的寄存器中。如果寄存器变量未存储在CPU的寄存器中，那么这个变量将变为默认的自动存储类型的变量。注意，许多编译器为了优化程序的执行速度，会将一些像循环索引变量这样的变量自动存储在CPU寄存器中。因此，多数情况下的循环索引变量可能自动指定为寄存器变量，而不必人为指定。

11.10 Command line arguments（命令行参数）

Many programs can be executed by typing the program name followed by a list of values that are passed to the program. For example, in the Linux operating system, the `cp` command is used to copy a file:

很多程序可以通过指定程序名和紧随其后的要传给程序的一个数据列表来执行。例如在Linux操作系统中，我们可以使用cp命令来实现文件的复制。

```
cp old_file new_file
```

In this example, `new_file` and `old_file` are known as *command line arguments*.

old_file和new_file称为**命令行参数**。

The number of command line arguments varies with the command. For example, the following `ls` command uses just one command line argument to get a directory listing of all text files:

```
ls *.txt
```

The operating system lets a C program know what is on the command line by passing it two pieces of information. The first piece of information is the number of arguments, and the second piece of information is the address of an array where the argument values are stored.

The next program shows how to access and display command line arguments.

操作系统通过向程序传递两个信息来告知C程序命令行中有哪些内容，第一个信息是命令行参数的个数，第二个信息是存储命令行参数的数组的首地址。

Program Example P11L

```
1  /* Program Example P11L
2     Demonstration of command line arguments. */
3  #include <stdio.h>
4  int main( int argc, char *argv[] )
5  /* argc is the number of arguments.
6     argv is an array of pointers to the arguments. */
7 {
8    int i ;
9
10   printf( "The number of arguments passed is %d\n\n", argc ) ;
11   printf( "The arguments passed are:\n\n" ) ;
12   for ( i = 1 ; i < argc ; i++ )
13     printf( "Argument %d is %s\n\n", i, argv[i] ) ;
14   return 0 ;
15 }
```

Note how `main()` is passed values from the operating system: the operating system treats `main()` as a function. Just as in any function, the passed values are stored in two parameters: `argc` (argument count) and `argv` (argument vector). The parameter names can be any valid variable names, but by convention they are named `argc` and `argv`.

Line 4 of this program declares `argc` as an integer and `argv` as an array of pointers to character strings. Each element of `argv` contains an address of the first character of a command line argument.

Run this program by typing the program name and some arguments. For example:

```
P11L hello out there
```

The output from this program will be:

```
The number of arguments passed is 4
The arguments passed are:
Argument 1 is hello
Argument 2 is out
Argument 3 is there
```

The program name is counted as the first argument, so `argc` will always be at least 1. This is why the `for` loop in line 12 starts at 1 and not `0`. You can display the program name by modifying the `for` loop in line 12 to start at `0`.

请注意main( )函数是如何接收操作系统传来的参数值的：操作系统将main( )当做一个普通函数，像其他函数一样，操作系统传给main( )的数据值存储在argc（参数个数argument count 的缩写）和argv（参数向量argument vector的缩写）这两个形参中，这两个形参名也可以是其他任何合法的变量名，但是，习惯上将它们命名为argc和argv。

程序的第4行，将argc定义为整型变量，将argv定义为字符指针数组，数组argv的每个元素中存储着各个命令行参数的首地址。

11.11 Mathematical functions（数学函数）

To use any of the mathematical functions place the preprocessor directive

```
#include<math.h>
```

at the start of your program.

使用C语言提供的数学函数时，必须在程序开头加上编译预处理指令。

11.11.1 Some commonly used trigonometric functions

Table 11.1 below lists the three common trigonometric functions sine, cosine and tangent.

表11.1列出了三种常用的三角函数：正弦、余弦和正切。

Table 11.1 common trigonometric functions

Function	Description
`cos(x)`	Cosine of angle `x` in radians. `x` is a `double` value. Returns a `double` value.
`sin(x)`	Sine of angle `x` in radians. `x` is a `double` value. Returns a `double` value.
`tan(x)`	Tangent of angle `x` in radians. `x` is a `double` value. Returns a `double` value.

The next program displays the `sin()`, `cos()` and `tan()` of an angle in degrees. The angle may be typed on the command line or input during the execution of the program.

下面的程序将在屏幕上输出某角度值的sin( )、cos( )、tan( )值，该角度值既可以用命令行参数的方式给出，也可以在程序运行时从键盘输入。

Program Example P11M

```
1  /* Program Example P11M
2     Demonstration of the functions sin(), cos(), and tan().
3     The angle (in degrees) may be typed on the command line or
4     input when the program is running.                          */
5  #include <stdio.h>
6  #include <stdlib.h>
7  #include <math.h>
8  #define DEGREES_TO_RADIANS 3.14157/180
9  int main( int argc, char **argv )
10 {
11   double degrees, radians ;
12
13   /* Has the user typed the angle on the command line? */
14   if ( argc < 2 )
15   {
16     /* No value on command line so ask the user for the angle. */
17     printf( "Input the angle in degrees: " ) ;
18     scanf( "%lf", &degrees ) ;
19   }
20   else
21   {
22     /* The user typed the angle on the command line.
23        The angle must be converted from an ASCII string
24        to a floating-point value using atof().                */
25     degrees = atof( argv[1] ) ;
```

```
26     }
27     /* Convert degrees to radians. */
28     radians = degrees * DEGREES_TO_RADIANS ;
29     /* Display the results. */
30     printf( "sin(%.3lf)= %.3lf\n", degrees, sin(radians) ) ;
31     printf( "cos(%.3lf)= %.3lf\n", degrees, cos(radians) ) ;
32     printf( "tan(%.3lf)= %.3lf\n", degrees, tan(radians) ) ;
33     return 0 ;
34 }
```

Run this program by typing the program name and the angle on the same line:

```
P11M 60
```

Alternatively, type the program name only

```
P11M
```

and you will be prompted for the angle:

```
Input the angle in degrees: 60
```

Both ways of running the program will display the same result:

```
sin(60.000)= 0.866
cos(60.000)= 0.500
tan(60.000)= 1.732
```

11.11.2 Other common mathematical functions

Table 11.2 below lists the more common mathematical functions that may be of use or that you may come across in other programs.

Table 11.2 common mathematical functions

Function	Description
`abs(n)`	Absolute value of an integer `n`. Returns an integer value.
`exp(x)`	Computes e^x. `x` is a `double` value. Returns a `double` value.
`fabs(x)`	Computes the absolute value of a floating-point number. `x` is a `double` value. Returns a `double` value.
`fmod(x,y)`	Computes the floating-point remainder of `x/y`. `x` and `y` are `double` values. Returns a `double` value.
`log(x)`	Computes the natural log of `x`. `x` is a `double` value > 0. Returns a `double` value.
`log10(x)`	Computes the base 10 log of `x`. `x` is a `double` value > 0. Returns a `double` value.
`pow(x,y)`	Computes x^y. `x` and `y` are `double` values. Returns a `double` value.
`sqrt(x)`	Computes the square root. `x` is a `double` value that is greater than or equal to 0. Returns a `double` value.

11.11.3 Pseudo-random number functions

To use the pseudo-random generating functions `rand()` and `srand()` (see Table 11.3), place the preprocessor directive

```
#include <stdlib.h>
```

at the start of your program.

使用C语言提供的伪随机数函数rand( )和srand( )时，必须在程序开头加上编译预处理指令。

Table 11.3 random number functions

Function	Description
`rand()`	Returns a pseudo-random integer value. Each call to `rand()` will produce a pseudo-random integer value. However, each time the program is executed the same sequence of integer values will be returned, unless a different *seed* value is used with `srand()`.
`srand(n)`	This function is used to set the seed (starting value) for the pseudo-random numbers generated by `rand()`. The seed value, n, is any positive integer value.

11.11.4 Some time-related functions

The time-related functions(see Table 11.4) can be used by placing the preprocessor directive

```
#include <time.h>
```

at the start of your program.

使用C语言提供的时间相关函数时，必须在程序开头加上编译预处理指令。

Table 11.4 time functions

Function	Description
`t=time(NULL)`	Returns to `t` the current time (measured in seconds since midnight on 1 January 1970, GMT).
`ctime(&t)`	Converts the time t returned by `time()` to a character string containing the current date and time. Returns a pointer to the character string containing the date and time.

The next program demonstrates the use of the time and random functions.

Program Example P11N

```
1  /* Program Example P11N
2     To demonstrate the built-in time and random functions. */
3  #include <stdio.h>
4  #include <stdlib.h>
5  #include <time.h>
6  int main()
7  {
8    int i, r ;
9    time_t t ;  /* Define t as being a time type time_t,
10                  as defined in time.h                   */
11
12   /* Display the current date and time. */
13   t = time( NULL ) ;
14   printf( "The current date and time are: %s\n", ctime( &t ) ) ;
15
16   /* Use the time to initialise the random number generator. */
17   srand( t ) ; /* Set the seed to the current time. */
18
```

```
19     /* Generate five random numbers between 0 and 20. */
20     printf( "Five random numbers in the range 0-20:\n" ) ;
21     for ( i = 0 ; i < 5 ; i++ )
22     {
23       r = rand() % 21 ;  /* %21 ensures a number between 0 and 20. */
24       printf( "%d\n", r ) ;
25     }
26     return 0 ;
27 }
```

A sample run of this program is:

```
The current date and time are: Fri Oct 21 12:08:08 2016

Five random numbers in the range 0-20:
11
8
17
9
3
```

Without line 17, the program displays the same sequence of "random" numbers every time the program is run. That's why they are called "pseudo-random" numbers – they are generated by the computer and are not truly random numbers. Try deleting line 17 and see what happens.

Line 17 uses the time as the seed (starting value), thereby generating a different set of "random" numbers every time the program is run.

如果没有第17行的语句，那么每次运行程序都将输出相同的"随机"数序列。由于这些随机数是由计算机生成的、并非真正的随机数，因此将其称为伪随机数。我们可以将第17行语句删除，从而观察程序的输出结果会发生什么变化。

程序第17行语句将系统当前时间作为随机数种子（初始值），因此，程序每次运行都将生成不同的随机数序列。

11.12 Recursion（递归）

Recursion is a programming technique in which a problem can be defined in terms of itself.

The technique involves solving a problem by reducing the problem to smaller versions of itself. For example, the factorial of a positive integer is the product of the integers from 1 through to that number. For example, 3 factorial (written as 3!) is 3*2*1 which is also 3*2!

The mathematical definition of factorial of a number n is:

$$n! \text{ is } \begin{cases} 1 \text{ when n is } 0 \\ n * (n-1)! \text{ when } n>0 \end{cases}$$

From this definition,

(a) 0!=1. This is called the *base case*.

(b) For a positive integer n, factorial n is n times the factorial of n−1. This is called the *general case* and clearly indicates that factorial is defined in terms of itself and is therefore an example of recursion.

Using the definition, factorial 3 is calculated as follows:

递归是一种根据问题本身来定义问题的编程技术，它通过将问题分解为与其自身相同的、只是规模较小的子问题来解决问题。

例如，一个正整数的阶乘就是从1到该正整数之间的所有正整数的乘积，例如，3的阶乘（记为3!）是3*2*1，也可以写成3*2!。

n的阶乘的数学定义如下。

(a) 0! = 1，称为**基线情况**。

(b) 正整数n的阶乘是n乘以n−1的阶乘，称为**一般情况**。显然这表明阶乘是根据其自身来定义的问题，因此它是一个典型的递归的例子。

- The value of n is 3 so, using (b) above, 3! = 3 * 2!
  - Next find 2! Here n = 2 so, using (b) again, 2! = 2 * 1!
    - Next find 1! Here n = 1 so, using (b) again, 1! = 1 * 0!
      - ☆ Next find 0! In this case using (a), 0! is defined as 1.
    - Substituting for 0! gives 1! = 1 * 1 = 1
  - Substituting for 1! gives 2! = 2 * 1! = 2 * 1 = 2.
- Finally, substituting for 2! gives 3! = 3 * 2! = 3 * 2 = 6.

The next program calculates the factorial of any positive integer.

Program Example P11O

```
1  /* Program Example P11O
2     Computation of the factorial of a number using recursion. */
3 #include <stdio.h>
4 unsigned int factorial( int n ) ;
5 int main()
6 {
7   unsigned int fact_n ;
8   int n ;
9
10   do  /* Read a valid number from the keyboard. */
11   {
12     printf( "Enter zero or a positive number " ) ;
13     scanf("%d", &n ) ;
14   }
15   while ( n < 0 ) ;
16
17   fact_n = factorial( n ) ;
18   printf("Factorial %d is %d\n", n , fact_n) ;
19
20   return 0 ;
21 }
22
23 /* Function : factorial
24    Purpose  : Recursive function to calculate n!
25    Parameter: The number for which the factorial is required.
26    Returns  : n!                                           */
27 unsigned int factorial( int n )
28 {
29   if ( n == 0 )
30     return 1 ;  /* Base case. */
31   else
32     return ( n * factorial(n-1) ) ; /* Function calls itself. */
33 }
```

A sample run of this program follows.

```
Enter zero or a positive number 5
Factorial 5 is 120
```

Note that

- Every recursive function must have at least one base case which stops the recursion and
- a general case that eventually reduces to a base case.

注意：

- 每一个递归函数必须至少有一个基线情况，它是用来结束递归过程的。
- 一般情况必须最终能简化为基线情况。

The factorial function could also be written using iteration, i.e. using a loop.

```
/* Function  : factorial
   Purpose   : Iterative function to calculate n!
   Parameter : The number for which the factorial is required.
   Returns   : n!                                               */
unsigned int factorial( int n )
{
  unsigned int fact ;
  int i ;
  fact = 1 ;
  for ( i = 2 ; i<= n ; i++ )
    fact *= i ;
  return fact ;
}
```

The recursive version will execute more slowly than the iterative equivalent because of the added overhead of the function calls in line 32 of program P11O.

The advantage of the recursive version, on the other hand, is that it is clearer because it follows the actual mathematical definition of factorial.

与用迭代方法编写的程序相比，用递归方法编写的程序的执行效率较低，这是因为增加了程序P11O第32行函数递归调用的开销。不过，由于递归程序遵循了阶乘的实际数学定义，因此用递归方法编写的程序具有可读性更好的优点。

Programming pitfalls

1. Parameters and arguments must agree in number and type. For example, if you call a function with two ints and a float as arguments, then you must have two ints and a float as parameters.
2. Be aware of the difference between passing arguments by value and passing arguments by reference. When you pass an argument by value, the value of that argument cannot be changed from within the function. To change the value of an argument from within a function you must pass the address of that argument to the function.
3. Parentheses ( ) are used to enclose function arguments; brackets [ ] are used for the subscripts of an array.
4. There is no semicolon after a function header, but there is one after a function prototype.
5. The angle in the trigonometric and hyperbolic functions is measured in radians, not degrees.
6. Leaving out the extern keyword will make a global variable declared within a function into an auto variable. For example, if line 23 of program P11K is changed from

```
extern int company_number ;
```

to

```
int company_number ;
```

then company_number would be regarded as a new local variable, distinct from the global variable of the same name.

7. Leaving out the data type of an external variable defaults that variable to type int. For example:

```
extern v ; /* v is, by default, an integer. */
```

8. The following function is meant to return a pointer to an integer:

```
int *funct()
{
  int int_val ;
  ...
  return &int_val;
}
```

This will not work. The problem here is that by the time the function returns to the function that called it, the integer variable int_val no longer exists. This is because the variable int_val is an auto variable, which means that the storage it occupied is automatically freed when the function returns.

9. If you omit a function prototype, C assumes the function will return an integer value. If the function is returning a data type other than

1. 函数的形参与实参必须在数量上和类型上一致。例如，如果希望用两个整数和一个浮点数作为函数实参去调用一个函数，那么必须将函数形参的类型声明为两个整型和一个浮点类型。
2. 注意按值传参和按引用传参的区别。按值传参时，实参的值是不能在函数内部修改的，若要在函数内部修改实参的值，那么必须向函数传递该实参变量的地址值。
3. 必须使用括号( )将函数参数括起来，而方括号[ ]用于将数组的下标括起来。
4. 在函数定义头部的末尾是没有分号的，而在函数原型的后面则需加上分号。
5. 在C语言提供的三角函数和反三角函数中，使用的单位为弧度，而非角度。
6. 对于在函数内部声明的全局变量，如果删掉其前面的关键字extern，那么系统将默认该变量的存储类型为自动存储类型。
7. 如果将全局变量声明中的数据类型删除，那么系统将默认该变量为整型。
8. 下面的函数试图返回一个整型指针。

变量int_val是一个自动变量，在函数调用结束返回时，其占用的存储空间已被系统自动释放了。

9. 如果在程序中省略函数原型，那么系统默认该函数

integer, the compiler will give an error message such as:

```
type mismatch with previous implicit declaration of
function
```

or

```
type mismatch in redeclaration of function
```

Avoid these messages by always including a prototype for every function in your programs.

的返回值类型为整型，如果该函数实际返回的数据类型不是整型，那么编译器将给出相关出错信息。为了避免此种情况的发生，必须对程序中的每个函数都给出函数原型。

Quick syntax reference

	Syntax	Examples
Function prototype	type function_name (type parameter$_1$, type parameter$_2$, ... type parameter$_n$) ;	float average (float a[], int n) ;

Function definition	type function_name
	(type parameter$_1$,
	type parameter$_2$,
	...
	type parameter$_n$)
	{
	local variables ;
	executable statements ;
	return expression ;
	}
		float average
		(float a[], int n)
		{
		int i ;
		float sum=0, average ;
		for (i=0; i<n; i++)
		sum +=a[i] ;
		average = sum/n ;
		return average ;
		}

Function call	variable = function_name(argument$_1$, argument$_2$, ... argument$_n$) ;	float array[10], n ; float avg ; ... avg=average(array, n) ;

Exercises

1. (a) Write function prototypes for the following functions:

Function	*parameter(s)*	*Return value*
f1	int	char
f2	int	int
f3	two ints	none
f4	pointer to char	pointer to char
f5	none	float
f6	none	none

For example, the function prototype for the function `f1()` is:

```
char f1( int variable_name ) ;        or     char f1( int ) ;
```

(b) Write a statement to call each of the above functions.

For example, to call function `f1()`:

```
char c_val ;
int i_val ;
c_val = f1( i_val ) ;
```

2. What is wrong with each of the following functions?

(a)
```
void max(a,b) ;
{
int a, b ;
if ( a > b )
  return a ;
else
  return b ;
```

(b)
```
void test(int)
{
  int i ;
  for(i=1;i< n;i++)
    putchar("x") ;
}
```

(c)
```
float min()
int a, b ;
if( a < b )
  return
return b ;
}
```

3. Find the errors in this program:

```
#include <stdio.h>
int main()
{
  int f1( void ) ;
  void f2( int ) ;
  int v1 ;
  float v2 ;
  double v3 ;
  v1 = f1[1] ;
  v2 = f1( ) ;
  v3 = f2( ) ;
  v2 = f2( 1, 2, 3 ) ;
    return 0 ;
  }

void f1( int a )
{
  /* Statements in function f1. */
}
void f2( int a, b )
{
  /* Statements in function f2. */
}
```

4. What is the output from the following program?

```
#include <stdio.h>
int main()
{
  int f( int val1, int val2 ) ;
  int var ;
  var = f( 1, 2 ) + 1 ;
  var = f( var + 1, 2 ) ;
  var = f( f( 1, 2 ), f( 3, var ) ) ;
  printf( "The value of var is %d", var ) ;
  return 0 ;
}
```

```
int f( int val1, int val2 )
{
  if ( val1 > val2 )
    return val1 - val2 ;
  else
    return val2 - val1 ;
}
```

5. Write a function to return the middle value of three integer values.
6. (a) Write a function to return the minimum value in an integer array of ten elements.
   (b) Modify the function to take the number of elements in the array as an argument.
7. Write a function to test whether an integer lies within a range of values. The function prototype will be:

   ```
   int range_test( int val, int low, int high ) ;
   ```

   where `val` is the value to be tested, `low` is the lower value in the range, and `high` is the higher value in the range. The function will return 1 if the value is in the specified range, otherwise it will return `0`.
8. Write a function to convert hours, minutes and seconds to seconds.
9. Write a function to convert seconds to hours, minutes, and seconds.
10. Write your own version of the standard library function `strlen()`.
    Name your function `my_strlen()`.
11. Write a function `rtrim()` to remove trailing spaces from the end of a string.
12. Write a function to count the number of words in a string.
    (Assume that every word is separated by at least one whitespace character.)
13. Write a function to capitalise all the letters in a string.
14. Write a function to determine the frequency of each of the vowels in a piece of text.
15. Write a function to reverse a string.
16. A palindrome is a word that you can spell forwards or backwards. For example, the words "madam" and "rotavator" are palindromes. Write a function which returns `1` if a word is a palindrome or returns `0` if it is not.
17. Write a function to determine whether a string is a palindrome or not. Examples of palindromic strings are "straw warts" and "rats live on no evil star".
    The function returns `1` if a string is a palindrome or `0` if it is not.
18. (a) What does the following program display?

    ```
    #include <stdio.h>
    void any_function( void ) ;
    int main()
    {
      int i ;
      for ( i = 0 ; i < 10 ; i++ )
        any_function() ;
      return 0 ;
    }

    void any_function()
    {
      static int var = 10 ;
    ```

```
    printf( "%d\n", var ) ;
    var += 10 ;
  }
```

(b) What would be the effect of replacing the function `any_function()` with the following:

```
void any_function()
{
  static int var ;
  var = 10 ;
  printf( "%d\n", var ) ;
  var += 10 ;
}
```

19. Write a function that will return the letter `A` the first time it is called, `B` the second time it is called, `C` the third time it is called, and so on. (Hint: use a `static` data type.)
20. Write a program to generate a set of lottery numbers in the range 1 to 42, using `srand()` and `rand()`.
21. Write a function that has an integer parameter and returns the leading digit of that integer. For example, the leading digit of 123 is 1 and the leading digit of –123 is –1.
22. What does this recursive function do?

```
void recur_fun( int n )
{
  printf( "%d\n", n ) ;
  if ( n == 1)
    return ;
  recur_fun ( n - 1 ) ;
}
```

Chapter Twelve Structures
第 12 章 结 构 体

The items of information that make up an array are all of the same data type (`int`s, `float`s etc.) and are logically related in some way. For example, a student's test scores may be integer values that are logically related to the student. In this case it makes sense to store the test scores together in an array. In short, arrays are suitable for storing sets of homogeneous data.

构成数组的所有信息项都具有相同的数据类型（如整型、浮点型等），并且在某种程度上是逻辑相关的。例如，学生的考试成绩是整型数据，并且这些数据与学生对象是逻辑相关的。在这种情况下，将所有的考试成绩存储在一个数组中才是有意义的，简而言之，数组适合存储具有相同类型的数据集合。

Sometimes items of information that are logically related are not of the same data type. A student's name and test scores, for example, are logically related to the student, but the name is a character string, while the test scores may be integers. C allows you to logically relate items of information that may be of different data types by combining the items into a *structure*. Unlike an array, the data items in a structure may have different data types.

有时，某些信息虽然是逻辑上相关的，但是数据的类型却不一致。例如，学生的姓名和考试成绩均与学生对象逻辑相关，但是，学生姓名是字符串，而考试成绩则可能是整型的。C语言允许将一组逻辑相关但类型不一致的一组数据信息组合在一起，形成一个**结构体**。与数组不同的是，结构体中的数据项可能具有不同的数据类型。

12.1 Defining a structure（定义结构体）

The first step in defining a structure is to declare a *structure template*:

定义结构体的第一步是为该结构体声明一个**结构体模板**。

```
struct student_rec
{
  int number ;            /* Student number.                 */
  char surname[21] ;      /* Surname is 20 characters long.  */
  char first_name[11] ;   /* First name is 10 characters long.*/
  int scores[5] ;         /* Scores on five tests.           */
} ;
```

A structure template consists of the reserved keyword **struct** followed by the name of the structure. The name of the structure is known as the *structure tag*. In the example above, `student_rec` is a structure tag.

结构体模板由保留的关键字struct和紧随其后的结构体名组成，结构体名也称为**结构体标签**。在上例中，student_rec就是一个结构体标签。

After the structure tag, each item within the structure is declared within the braces `{` and `}`. Each item in a structure is called a *structure member*. A structure member has a

在结构体标签后面的一对花括号中，声明了结构体内的每个数据项，结构体内的数据项称为**结构体成员**。每个结构体成员都有它的

name and a data type. You can choose any name you wish for a member, provided you obey the rules for constructing valid variable names (see Chaper Two).

名字和数据类型，成员的名字可以是任意的，只要遵循变量的命名规则，是合法的变量名即可。

Declaring a structure template does not allocate memory to the structure. All that has been done at this stage is to define a new data type consisting of other previously defined data types. Once you have defined the new data type you can then define variables with that type. For example,

声明结构体模板时，并不为结构体分配内存。在这一步所做的工作只是定义了一种新的数据类型，这种数据类型由其他以前已定义的数据类型构成。这种新的数据类型一旦定义好以后，就可以用它来定义具有此种数据类型的变量了。

```
struct student_rec student1, student2 ;
```

defines the variables `student1` and `student2` to be of the type `struct student_rec`. Both `student1` and `student2` are structure variables with four members, i.e. `number`, `surname`, `first_name`, and `scores` as shown in Figure 12.1.

```
student1:
┌──────────┐
│  number  │
├──────────┴──────────────────────────────────────────────┐
│                        surname                          │
├─────────────────────┬───────────────────────────────────┤
│     first_name      │                                   │
├──────────┬──────────┼──────────┬──────────┬─────────────┤
│scores[0] │scores[1] │scores[2] │scores[3] │scores[4]    │
└──────────┴──────────┴──────────┴──────────┴─────────────┘
student2:
┌──────────┐
│  number  │
├──────────┴──────────────────────────────────────────────┐
│                        surname                          │
├─────────────────────┬───────────────────────────────────┤
│     first_name      │                                   │
├──────────┬──────────┼──────────┬──────────┬─────────────┤
│scores[0] │scores[1] │scores[2] │scores[3] │scores[4]    │
└──────────┴──────────┴──────────┴──────────┴─────────────┘
```

Figure 12.1 `student1` and `student2` structure variables

The members of a structure variable can be accessed with the member selection operator "." (a dot). For example, we can assign values to the member `number` of the variables `student1` and `student2` with the statements:

可以使用成员选择运算符“.”来访问结构体变量的每个成员。

```
student1.number = 1234 ;        student2.number = 13731 ;
```

You can use `student1.number` and `student2.number` in the same way that you use any other integer variable.

The next program inputs values for each member of a structure variable and displays them on the screen.

可以用与其他整型变量相同的方式来使用结构体的整型成员变量student1.number和student2.number。

Program Example P12A

```
1   /* Program Example P12A
2      This program assigns values to two structure variables. */
3   #include <string.h>
4   #include <stdio.h>
5   int main()
6   {
7     /* Declare the structure template. */
8     struct student_rec
9     {
10      /* Declare the members of the structure. */
11      int number ;
```

```
12      char surname[21] ;
13      char first_name[11] ;
14      int scores[5] ;
15    } ;
16
17    /* Define two variables having the type struct student_rec. */
18    struct student_rec student1, student2 ;
19
20    int i ;
21
22    /* Read in values for the members of student1. */
23    printf( "Number: " ) ;
24    scanf( "%d", &student1.number ) ;
25    printf( "Surname: " ) ;
26    scanf( "%20s", student1.surname ) ;
27    printf( "First name: " ) ;
28    scanf( "%10s", student1.first_name ) ;
29    printf( "Five test scores: " ) ;
30    for ( i= 0 ; i < 5 ; i++ )
31      scanf( "%d", &student1.scores[i] ) ;
32
33    /* Now assign values to the members of student2.
34       The assignments are not meant to be meaningful: they
35       are for demonstration purposes only.                */
36
37    student2.number = student1.number + 1 ;
38    strcpy( student2.surname, "Smith" ) ;
39    strcpy( student2.first_name, "Mary" ) ;
40    for ( i= 0 ; i < 5 ; i++ )
41      student2.scores[i] = 100 ;
42
43    /* Display the values in the members of student1. */
44    printf( "\n\nThe values in student1 are:" ) ;
45    printf( "\nNumber is %d", student1.number ) ;
46    printf( "\nSurname is %s", student1.surname ) ;
47    printf( "\nFirst name is %s", student1.first_name ) ;
48    printf( "\nScores are: " ) ;
49    for ( i= 0 ; i < 5 ; i++ )
50      printf( " %d ", student1.scores[i] ) ;
51
52    /* Display the values in the members of student2. */
53    printf( "\n\nThe values in student2 are:" ) ;
54    printf( "\nNumber is %d", student2.number ) ;
55    printf( "\nSurname is %s", student2.surname ) ;
56    printf( "\nFirst name is %s", student2.first_name ) ;
57    printf( "\nScores are: " ) ;
58    for ( i= 0 ; i < 5 ; i++ )
59      printf( " %d ", student2.scores[i] ) ;
60    printf( "\n" ) ;
61    return 0 ;
62  }
```

Running this program produces the following:

```
Number: 1234
Surname: Li
First name: Dong
Five test scores: 50 60 45 65 75

The values in student1 are:
Number is 1234
Surname is Li
First name is Dong
Scores are: 50  60  45  65  75

The values in student2 are:
Number is 1235
Surname is Smith
First name is Mary
Scores are: 100  100  100  100  100
```

The structure tag `student_rec` in line 8 of program P12A is optional when the structure template and the structure variables are defined together:

同时定义结构体模板和结构体变量时，程序P12A第8行的结构体标签student_rec可以省略。

```
/* Declaring a structure template without a structure tag. */
struct    /* No tag name after struct. */
{
  int number ;
  char surname[21] ;
  char first_name[11] ;
  int scores[5] ;
} student1, student2 ; /* The variables are immediately after }. */
```

Like other variable types, a structure variable can be assigned to another structure variable of the same type. For example,

与其他类型的变量类似，可以将一个结构体变量赋值给另一个相同类型的结构体变量。

```
student2 = student1 ;
```

assigns each member of `student2` the value of the corresponding member of `student1`.

Try modifying program P12A so that `student2` is identical to the input values of `student1`.

12.2 Pointers to structures（结构体指针）

The general format for defining a pointer to a structure is:

```
struct tag_name *variable_name ;
```

where `tag_name` is the structure tag and `variable_name` is the name of the pointer variable. For example, the following line defines a pointer `ptr` to the `student_rec` structure in program P12A:

```
struct student_rec *ptr ;
```

You can assign `ptr` a value by using the address operator `&`, as in:

可以使用取地址运算符&来为指针变量ptr赋值。

注意，这里用结构体变量student1（而非结构体标签

```
ptr = &student1 ;
```

Note that you assign the address of the structure variable `student1` and not the address of the structure tag `student_rec`.

student_rec）的地址来为指针变量赋值。

You can refer to the members of a structure variable by using the dereferencing operator *. For example,

同样可以使用解引用运算符*来引用结构体变量的成员。

```
(*ptr).number
```

will access the student's number. The parentheses are necessary, because the selection operator . has a higher priority than the dereferencing operator *. Without the parentheses (as in `*ptr.number`) you are attempting to access the memory location given by `ptr.number`. This is invalid, because `ptr` is not a structure and `number` is not a member of `ptr`.

由于成员选择运算符的优先级要高于解引用运算符，因此这里的圆括号是必须的。如果省略了圆括号（即*ptr.number），那么其含义将变成：试图访问ptr.number这个内存地址中的内容，由于ptr不是一个结构体变量，number也不是ptr的成员，因此ptr.number这种写法是非法的。

C provides us with a much more convenient notation for accessing the members of a structure that has a pointer to it. The arrow notation -> (- and > together) can be used in place of the dot notation. Thus,

```
ptr -> number    and    (*ptr).number
```

are equivalent. In English, `ptr->number` reads as "the member number of the structure pointed to by `ptr`".

C语言为我们提供了一种更为方便的记号，即用箭头标记->（-和>的组合）代替圆点标记，来访问结构体指针所指向的结构体的成员。因此，ptr->number与(*ptr).number二者是等价的，可将ptr->number读做“指针ptr指向的结构体的成员number”。

Program P12B demonstrates the use of a pointer to a structure variable, but before considering this program you will need to know how a structure variable can be initialised.

12.3 Initialising a structure variable（结构体变量的初始化）

The members of a structure variable can be initialised by placing the initial values between braces in the same way that array elements are initialised.

与数组元素的初始化相同，我们可以用放置在花括号内的初值对结构体变量的成员进行初始化。

Example:

```
struct student_rec
{
  int number ;
  char surname[21] ;
  char first_name[11] ;
  int scores[5] ;
} ;
struct student_rec student = { 1234,
                               "Li",
                               "Peiqi",
                               { 50, 60, 45, 65, 75 }
                             } ;
```

The first member of the structure (`student.number`) is initialised to `1234`, the second member (`student.surname`) is initialised to `"Li"`, and the third member (`student.first_name`) is initialised to

将结构体变量的初始值放在不同行中书写，只是为了可视化的目的，使用这种方式

"Peiqi". The last member (student.scores) is an integer array and is initialised to the values enclosed in the inner set of braces. The initial values are on separate lines for visual purposes only, so you can easily relate a structure member to its initial value.

可以很容易地将结构体的成员与其初始值联系起来。

The next program initialises a structure variable and uses a pointer to display the initial values.

Program Example P12B

```
1  /* Program Example P12B
2     Demonstration of structure initialisation and
3     structure pointers.                          */
4  #include <stdio.h>
5  int main()
6  {
7    struct student_rec     /* structure template. */
8    {
9      int number ;
10     char surname[21] ;
11     char first_name[11] ;
12     int scores[5] ;
13   } ;
14
15   /* Define a structure variable and initialise it with
16      some values. */
17   struct student_rec student = {
18                                1234,
19                                "Li",
20                                "Peiqi",
21                                { 50, 60, 45, 65, 75 }
22                              } ;
23   struct student_rec *ptr ;
24   int i ;
25
26   ptr = &student ; /* ptr points to student. */
27
28   /* Display the values in the members of student
29      using a pointer. */
30   printf( "The values in student are:" ) ;
31   printf( "\nNumber is %d", ptr -> number ) ;
32   printf( "\nSurname is %s", ptr -> surname ) ;
33   printf( "\nFirst name is %s", ptr -> first_name ) ;
34   printf( "\nScores are: " ) ;
35   for ( i= 0 ; i < 5 ; i++ )
36     printf( " %d ", ptr -> scores[i] ) ;
37   printf( "\n" ) ;
38   return 0 ;
39 }
```

When you run this program you will get the following output:

```
The values in student are:
Number is 1234
Surname is Li
First name is Peiqi
Scores are:  50  60  45  65  75
```

Lines 17 to 22 initialise the structure variable `student`. Line 26 assigns the address of `student` to the pointer variable `ptr`. Lines 31 to 33 and 36 use the pointer `ptr` to access the members of the structure.

12.4 Passing a structure to a function（向函数传递结构体变量）

When you pass a structure variable to a function, you pass a copy of the member values to that function. This means that the values in the structure variable cannot be changed within the function. As with any variable, to change the values in a structure variable from within a function you must pass to the function a pointer to the structure variable.

It is faster to pass a pointer to a structure than a copy of all the values in the structure variable. Passing a pointer is also more portable, as not all compilers support structure arguments.

The next program demonstrates the two techniques. The function `get_student_data()` uses a pointer to a structure variable, and the function `display_student_data()` uses the structure variable itself as an argument.

向函数传递结构体变量，实际上就是将结构体变量各个成员值的副本传递给函数。这就意味着结构体变量的值是无法在函数内部修改的。与其他类型的变量一样，若要在函数中修改结构体变量的值，那么必须向函数传递指向该结构体的指针。

向函数传递结构体指针比传递结构体变量各个成员值的副本要快得多，而且向函数传递结构体指针还增强了函数的可移植性，因为并不是所有的编译器都支持将结构体变量作为函数的实参。

Program Example P12C

```
1  /* Program Example P12C
2     To demonstrate passing a structure variable to a function.
3     Program reads and displays data for a student.           */
4  #include <stdio.h>
5
6  struct student_rec /* Global structure template. */
7  {
8    int number ;
9    char surname[21] ;
10   char first_name[11] ;
11   int scores[5] ;
12 } ;
13
14 void display_student_data( struct student_rec student );
15 void get_student_data( struct student_rec *ptr ) ;
16
17 int main()
18 {
19   struct student_rec student ;
20   struct student_rec *student_ptr ;
21
```

```
22   student_ptr = &student ;
23
24   /* Use a pointer to a structure variable as an argument. */
25   get_student_data( student_ptr ) ;
26
27   /* Use a structure variable as an argument. */
28   display_student_data( student ) ;
29   return 0 ;
30 }
31
32 /* Function : display_student_data()
33    Purpose  : This function displays student data.
34    Parameter: The student data.                      */
35 void display_student_data( struct student_rec student )
36 {
37   int i;
38   printf( "\nThe data in the student structure is:" ) ;
39   printf( "\nNumber is %d", student.number ) ;
40   printf( "\nSurname is %s", student.surname ) ;
41   printf( "\nFirst name is %s", student.first_name ) ;
42   printf( "\nScores are: " ) ;
43   for ( i= 0 ; i < 5 ; i++ )
44     printf( " %d ", student.scores[i] ) ;
45   printf( "\n" ) ;
46 }
47
48 /* Function : get_student_data()
49    Purpose  : This function reads student data.
50    Parameter: A pointer to the student data structure. */
51 void get_student_data( struct student_rec *ptr )
52 {
53   int i ;
54   printf( "Number: " ) ;
55   scanf( "%d", &(ptr->number) ) ;
56   printf( "Surname: " ) ;
57   scanf( "%20s", ptr->surname ) ;
58   printf( "First name: " ) ;
59   scanf( "%10s", ptr->first_name ) ;
60   printf( "Five test scores: " ) ;
61   for ( i= 0 ; i < 5 ; i++ )
62     scanf( "%d", &(ptr->scores[i]) ) ;
63 }
```

A sample run of this program follows.

```
Number: 1234
Surname: Li
First name: Dong
Five test scores: 50 60 45 65 75
The data in the student structure is:
Number is 1234
```

```
Surname is Li
First name is Dong
Scores are:  50  60  45  65  75
```

Lines 6 to 12 declare the structure template, as in previous programs. This time, however, the structure template is outside `main()`. When a structure template is defined outside `main()`, it makes the structure template *global*. This means that the structure template is known in `main()` and in the functions `display_student_data()` and `get_student_data()`. There is no need, therefore, to define the structure template in each function.

程序第6行~第12行声明了一个结构体模板，不过这里的结构体模板位于主函数之外。在主函数之外定义的结构体模板是全局的，这意味着该结构体模板不仅在主函数内可以使用，还可以在函数display_student_data( )和get_student_data( )内使用。这样，就不必在每个函数内都定义一个结构体模板了。

12.5 Nested structures（嵌套的结构体）

A *nested structure* is a structure that contains another structure as one of its members. For example, a company personnel record might consist of, among other things, the employees' date of birth and date of joining the company. Both these dates can be represented by a structure consisting of the members day, month, and year.

所谓**嵌套的结构体**是指在一个结构体中包含另一个结构体类型的成员。例如，记录公司人事档案的结构体可能包含的成员有：员工的出生日期、员工进入公司的日期及其他信息。而上述两个日期可以用一个包括年、月、日成员的结构体类型来描述。

First declare the structure template for a date as follows:

```
struct date  /* Structure template for a date. */
{
  int day ;
  int month ;
  int year ;
} ;
```

Next, the template for the structure `personnel` is declared in terms of the previously declared structure template `date`.

```
struct personnel  /* Structure template for an employee. */
{
  unsigned int number ;
  char surname[21] ;
  char first_name[11] ;
  struct date dob ;   /* dob is type struct date.          */
  int dept ;
  struct date joined ; /* joined is also of type struct date. */
} ;
```

Finally, define a variable `person` of the type `struct personnel`, as in:

```
struct personnel person ;
```

Graphically, the personnel structure is shown in Figure 12.2 below.

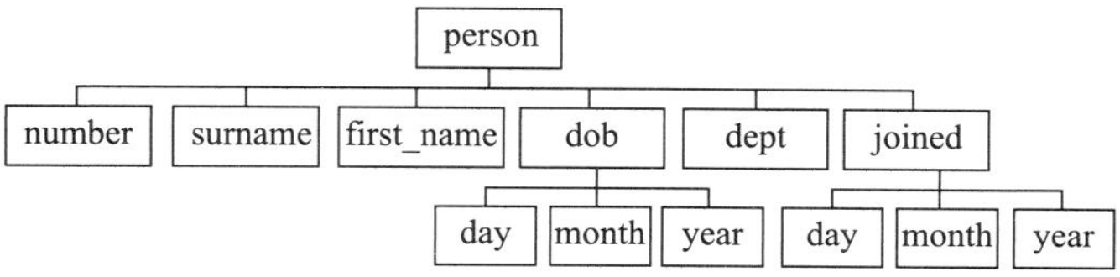

```
person
├── number
├── surname
├── first_name
├── dob
│   ├── day
│   ├── month
│   └── year
├── dept
└── joined
    ├── day
    ├── month
    └── year
```

Figure 12.2 the `personnel` structure

The expressions

```
person.dob
```

and

```
person.joined
```

will access the date of birth and date of joining members, respectively. Furthermore,

```
person.dob.day
person.dob.month
```

and

```
person.dob.year
```

will access the day, month and year of birth, respectively. Similarly,

```
person.joined.day
person.joined.month
```

and

```
person.joined.year
```

will access the day, month and year of the date the person joined the company.

The members of a nested structure can also be accessed by using a pointer. For example, if `ptr` is defined as

```
struct personnel *ptr = &person ;
```

then

```
ptr->dob.day
ptr->dob.month
ptr->dob.year
```

will access the date of birth members of the structure.

There is no limit to the level of nesting you can define. However, there is probably a practical limit of three or four levels, beyond which you can easily become confused.

C语言并未限制结构体嵌套的级数，但是，经验表明结构体的嵌套级数最好不要超过3级或者4级，因为级数太多，很容易把人搞糊涂。

12.6 Including a structure template from a file（从文件中引用结构体模板）

If a structure template is used in two or more programs, rather than typing the template into each program, you can use a *header file* that contains the structure template. A header file is simply a text file that typically contains C declarations and macro definitions (see Chapter Fourteen) to be shared between several programs. You request the use of a header file in your program by *including* it, with the `#include` preprocessor directive. For example, if the file `employee.h` contains the structure template for an employee, then the directive

如果一个结构体模板同时在两个或者更多的程序中使用，那么可以不用在每个程序中都重新定义结构体模板，我们可以将结构体模板的定义放在一个**头文件**中。头文件就是一个简单的文本文件，它包含了一些可以在多个程序间共享的C声明和宏定义

```
#include "employee.h"
```

will include the structure template in your program. The advantage of having the structure template in a header file is that you only have to type it once, but it can be included in as many programs as required.

（参见第14章）等。只要使用#include编译预处理指令将头文件包含到程序中，就可以使用这个头文件中的内容了。

12.7 The typedef statement （typedef 语句）

typedef allows you to define a synonym for an existing data type. For example, the statement

typedef允许为一个已有的数据类型定义一个别名。

```
typedef int * INT_POINTER ;
```

creates a synonym `INT_POINTER` for the data type `int*`. The synonym `INT_POINTER` can now be used in place of `int*` anywhere in a program. For example, instead of writing

```
int *p1, *p2, *p3 ;
```

you can now write

```
INT_POINTER p1, p2, p3 ;
```

Here is another example:

```
typedef char STRING ;
STRING name[21], address[31] ;
```

This is equivalent to:

```
char name[21], address[31] ;
```

By convention, a synonym name is written in capital letters, but this is not mandatory.

通常，习惯上用大写字母来命名数据类型的别名，但这不是强制性的要求。

You can also use `typedef` with structures. From the personnel example in section 12.5, we had the following structure templates:

也可以使用typedef为结构体类型定义别名。

```
struct date            /* Structure template for a date.      */
{
  int day ;
  int month ;
  int year ;
} ;

struct personnel       /* Structure template for an employee.  */
{
  unsigned int number ;
  char surname[21] ;
  char first_name[11] ;
  struct date dob ;    /* dob is of type struct date.          */
  int dept ;
  struct date joined ; /* joined is also of type struct date.  */
} ;
```

If you use

```
typedef struct date DATE ;
```

you can then define the personnel structure template as:

```
struct personnel
{
  unsigned int number ;
  char surname[21] ;
  char first_name[11] ;
  DATE dob ;
  int dept ;
  DATE joined ;
} ;
```

Going a step further, you can write

```
typedef struct personnel EMPLOYEE ;
```

and declare the variable `person` as:

```
EMPLOYEE person ;
```

12.8 Arrays of structures（结构体数组）

Continuing with the personnel example in section 12.5:

```
struct personnel persons[5] ;
```

or

```
EMPLOYEE persons[5] ;
```

defines a five-element array `persons`. Each element of this array is of the type `struct personnel`, with members `number`, `surname`, `first_name`, `dob`, `dept`, and `joined`. The members `dob` and `joined` are themselves structures and have members `day`, `month`, and `year`.

Note that `persons[0].number` will access the employee number of the first employee and `persons[4].joined.year` will access the year of joining of the fifth employee.

The next program maintains a database of employee data by using an array of structures. The data in this program is read from the keyboard and is stored in memory. However, the data is lost when the program finishes. This is a serious limitation for any real-life program. In Chapter Thirteen you will learn how to read and write data files that will allow you to store data permanently.

Run this program and study each function to familiarise yourself with the techniques used.

注意，persons[0].number访问的是persons数组中存储的第一个员工的员工编号，而persons[4].joined.year访问的是数组中存储的第5个员工加入公司的年份。

下面的程序使用结构体数组来维护一个人事数据库，程序中的数据是从键盘读入的，并且存储在内存中。不过，当程序结束时，数据就会丢失。对于有实用性的程序而言，这是一个严重的局限性。在第13章将学习如何从文件中读入数据，以及如何将数据写入文件，这样就可以永久地保存数据了。

Program Example P12D

```
1 /* Program Example P12D
2    This program maintains a database of employee records
3    in memory.
4    Written by    : Paul Kelly and Su Xiaohong.
5    Date          : 29/06/2012.
6 */
7
8 #include <stdio.h>
9
10 struct date
11 {
12   int day ;
13   int month ;
14   int year ;
15 } ;
16 typedef struct date DATE ;
17 struct personnel  /* Employee details. */
18 {
19   unsigned int number ;
20   char surname[21] ;
21   char first_name[11] ;
22   DATE dob ;
23   int dept ;
24   DATE joined ;
25 } ;
26 typedef struct personnel EMPLOYEE ;
27
28 void add_an_employee( EMPLOYEE person_array[] ) ;
29 void delete_an_employee( EMPLOYEE person_array[] ) ;
30 void display_an_employee( const EMPLOYEE person_array[] ) ;
31 void display_all_employees( const EMPLOYEE person_array[] ) ;
32 void display_employee_details( const EMPLOYEE *ptr ) ;
33 void explanation( void ) ;
34 void initialise_database( EMPLOYEE person_array[] ) ;
35 int menu( void ) ;
36 int search_database( const EMPLOYEE person_array[],
37                      const unsigned int emp_no ) ;
38
39 #define MAX_PERSONS 100
40
41 int main()
42 {
43   int menu_choice ;
44   EMPLOYEE persons[MAX_PERSONS] ; /* The database. */
45
46   explanation() ; /* Explain the purpose of the program.  */
47
48   initialise_database( persons ) ; /* Clear the database. */
49
```

```
50   do
51   {
52     menu_choice = menu() ; /* Get the user's choice */
53     switch ( menu_choice ) /* and act on it.       */
54     {
55       case 1 :
56         add_an_employee( persons ) ;
57         break ;
58       case 2 :
59         delete_an_employee( persons ) ;
60         break ;
61       case 3 :
62         display_an_employee( persons ) ;
63       case 4 :
64         display_all_employees( persons ) ;
65       }
66   }while ( menu_choice != 0) ;
67   return 0 ;
68 }
69
70 /* Function : add_an_employee()
71    Purpose  : Add an employee's data to the database.
72    Parameter: A pointer to the employee array.          */
73
74 void add_an_employee( EMPLOYEE person_array[] )
75 {
76   int i = 0 ;
77
78   /* Search the array for an empty position.
79      An empty position has a 0 for the employee number.  */
80   while ( person_array[i].number != 0 && i < MAX_PERSONS )
81     i++ ;
82   if ( i == MAX_PERSONS )
83     printf( "\nSorry, the database is full.\n" ) ;
84   else        /* Add the person's details to the database. */
85   {
86     printf( "\n\nEmployee number (must be greater than 0): " ) ;
87     do
88     {
89       scanf( "%5d", &person_array[i].number ) ;
90     } while ( person_array[i].number <= 0 ) ;
91
92     printf( "\n          Surname (maximum 25 characters): " ) ;
93     scanf( "%20s", person_array[i].surname ) ;
94     printf( "\n       First name (maximum 10 characters): " ) ;
95     scanf( "%10s", person_array[i].first_name ) ;
96     printf( "\nDate of birth\n" ) ;
97     printf( "     Day (1 or 2 digits): " ) ;
98     scanf( "%2d", &person_array[i].dob.day ) ;
99     printf( "   Month (1 or 2 digits): " ) ;
```

```
100     scanf( "%2d", &person_array[i].dob.month ) ;
101     printf( "    Year (1 or 2 digits): " ) ;
102     scanf( "%2d", &person_array[i].dob.year ) ;
103     printf( "\nDepartment code (1 to 4 digits): " ) ;
104     scanf( "%4d", &person_array[i].dept ) ;
105     printf( "\nDate joined\n" ) ;
106     printf( "     Day (1 or 2 digits): " ) ;
107     scanf( "%2d", &person_array[i].joined.day ) ;
108     printf( "    Month (1 or 2 digits): " ) ;
109     scanf( "%2d", &person_array[i].joined.month ) ;
110     printf( "    Year (1 or 2 digits): " ) ;
111     scanf( "%2d", &person_array[i].joined.year ) ;
112   }
113 }
114
115 /* Function : delete_an_employee()
116    Purpose  : Delete an employee from the database.
117    Parameter: A pointer to the employee array.          */
118
119 void delete_an_employee( EMPLOYEE person_array[] )
120 {
121   unsigned int emp_no ;
122   int pos ;
123
124   /* An employee is marked as 'deleted' from the database
125      by placing a 0 in the employee number.             */
126   /* First get the employee number. */
127   printf("Employee number to delete (must be greater than 0): ");
128   do
129   {
130     scanf( "%5d", &emp_no ) ;
131   }while ( emp_no <= 0 ) ;
132
133   /* Find the position of this employee in the database. */
134   pos = search_database( person_array, emp_no ) ;
135   /* Have you come to the end of the database without finding
136      the employee number? */
137   if ( pos == MAX_PERSONS ) /* yes */
138     printf( "This employee is not in the database.\n" ) ;
139   else                      /* no */
140   {
141     printf( "Employee %5d deleted.", emp_no ) ;
142     person_array[pos].number = 0 ;
143   }
144 }
145
146 /* Function : display_all_employees()
147    Purpose  : Display all employee's details in the database.
148    Parameter: A pointer to the employee array.            */
149
```

```
150 void display_all_employees( const EMPLOYEE person_array[] )
151 {
152   int i ;
153
154   for ( i = 0 ; i < MAX_PERSONS ; i++ )
155     if ( person_array[i].number != 0 )
156       display_employee_details( &person_array[i] ) ;
157 }
158
159 /* Function : display_an_employee()
160    Purpose  : Display an employee's details in the database.
161    Parameter: A pointer to the employee array.              */
162
163 void display_an_employee( const EMPLOYEE person_array[] )
164 {
165   unsigned int emp_no ;
166   int pos ;
167
168   /* Get the employee number. */
169   printf("Employee number to display (must be greater than 0):" );
170   do
171   {
172     scanf( "%5d", &emp_no ) ;
173   } while ( emp_no <= 0 ) ;
174
175   /* Find the position of this employee in the database. */
176   pos = search_database( person_array, emp_no ) ;
177   /* Does the employee exist? */
178   if ( pos == MAX_PERSONS )   /* no */
179     printf( "This employee is not in the database.\n" ) ;
180   else                        /* yes - display the details. */
181     display_employee_details( &person_array[pos] ) ;
182 }
183 /* Function : display_employee_details()
184    Purpose  : Display employee details.
185    Parameter: A pointer to the employee details.            */
186
187 void display_employee_details( const EMPLOYEE *ptr )
188 {
189   printf( "\n\nEmployee number: %d\n", ptr->number ) ;
190   printf( "        Surname: %s\n", ptr->surname) ;
191   printf( "    First name:  %s\n", ptr->first_name ) ;
192   printf( "  Date of birth: %2d/%2d/%2d\n",
193          ptr->dob.day, ptr->dob.month, ptr->dob.year ) ;
194   printf( "    Department: %d\n", ptr->dept) ;
195   printf( "  Date joined:  %2d/%2d/%2d\n",
196          ptr->joined.day, ptr->joined.month, ptr->joined.year ) ;
197 }
198
199 /* Function  : explanation()
```

```
200    Purpose   : Explain the purpose and operation of program.
201    Parameters: None.                                          */
202
203 void explanation()
204 {
205   int i ;
206
207   for ( i = 0 ; i < 24 ; i++ ) /* Clear the screen. */
208     printf( "\n" ) ;
209
210   printf( "This program allows you to store and retrieve data "
211           "on up to %d employees.\n\n"
212           "The data is held in memory and is therefore lost"
213           "when the program terminates.\n\n",  MAX_PERSONS ) ;
214 }
215
216 /* Function : initialise_database()
217    Purpose  : Initialise all employee numbers to 0.
218               An employee number of 0 is used to denote
219               an empty position in the array.
220    Parameter: A pointer to the employee array.                 */
221
222 void initialise_database( EMPLOYEE person_array[] )
223 {
224   int i ;
225
226   for ( i = 0 ; i < MAX_PERSONS ; i++ )
227     person_array[i].number = 0 ;
228 }
229
230 /* Function   : menu()
231    Purpose    : Display a menu of options and wait for
232                 the user to choose one.
233    Parameters: None.
234    Returns    : The menu choice.                               */
235
236 int menu()
237 {
238   int choice ;
239
240   /* Display the menu. */
241   printf( "\n\n 1. Add     an employee\n\n" ) ;
242   printf( " 2. Delete  an employee\n\n" ) ;
243   printf( " 3. Display an employee\n\n" ) ;
244   printf( " 4. Display all employees\n\n" ) ;
245   printf( " 0. Quit\n\n" ) ;
246   printf( "Please enter your choice (0 to 3): " ) ;
247
248   /* Get the option. */
249   do
```

```
250   {
251     scanf( "%d", &choice ) ;
252   }while ( choice <0 || choice > 4 ) ;
253
254   return choice;
255 }
256
257 /* Function   : search_database()
258    Purpose    : Search the database for an employee number.
259    Parameters: A pointer to the employee array and
260                an employee number.
261    Returns    : The array index corresponding to the position
262                in the database of the employee record.        */
263
264 int search_database( const EMPLOYEE person_array[],
265                      const unsigned int emp_no )
266 {
267   int i = 0 ;
268
269   /* Search for the employee number in the database, starting at
270      the first record and continuing until the employee number is
271      found or until the end of the database is reached.        */
272   while ( person_array[i].number != emp_no && i < MAX_PERSONS )
273     i++ ;
274
275   return i ;
276 }
```

12.9 Enumerated data types（枚举数据类型）

An *enumerated* data type is used to describe a set of integer values. For example:

枚举数据类型主要用于描述一组整型值。

```
enum response { no, yes, none } ;
enum response answer ;
```

These statements declare the enumerated data type `response` to have one of three possible values: `no`, `yes`, or `none`. The variable `answer` is defined as an enumerated variable of type `response`. This is similar to the way in which a structure template and a structure variable are defined. The identifiers enclosed in the braces `{` and `}` are integer constants. The first identifier (`no`) has a value of 0, the second (`yes`) has a value of 1, and the third (`none`) has a value of 2. The variable `answer` can be assigned any of the possible values: `no`, `yes`, or `none`.
For example:

上述语句声明了一个枚举数据类型response，它有三个可能的取值：yes，no，none。变量answer定义为枚举类型response，可以看出，枚举类型和枚举变量的这种定义方式与结构体模板和结构体变量的定义是相似的。

```
answer = none ;
```

or

```
answer = no ;
```

The variable `answer` can also be used in an `if` statement. For example:

```
if ( answer == yes )
```

The purpose of the enumerated data type is to improve the readability of the program. In the example above, using `yes`, `no` and `none` rather than 0, 1 and 2 makes the program more readable.
In this example, `response` is called the *enumeration tag*. Like a structure tag, the enumeration tag is optional when the enumerated data type and the enumerated variables are defined together.
For example, the variable `answer` could also be defined as:

```
enum { no, yes, none } answer ;
```

使用枚举数据类型的目的是提高程序的可读性，在上面的例子中，我们使用yes、no和none比使用0、1、2的程序可读性更高。
上例中的response称为**枚举标签**，与结构体标签类似，同时定义枚举类型和枚举变量时，枚举标签是可以省略的。

Arrays can also be used. For example:

```
enum response answers[200] ;
```

You are not restricted to the default values 0, 1, and 2. You can specify any values you wish by including the required values in the braces. For example:

```
enum response { no = -1, yes = 1, none = 0 } ;
```

使用枚举类型时，枚举类型括号内的数值不只局限于默认的0、1、2，可以通过在括号内包含所需的数值来指定想要的数值。

If you want to add another possible value to `response`, include the new value within the braces. For example:

```
enum response { no = -1, yes = 1, none = 0, unsure = 2 } ;
```

Programming pitfalls

1. You cannot compare structure variables in an if statement, even if they have the same structure template. For example, if s1 and s2 are defined as

```
struct
{
  int a ;
  int b ;
  float c ;
} s1, s2 ;
```

you cannot test s1 and s2 for equality with the statement:

```
if ( s1 == s2 )    /* Invalid! */
```

To test s1 and s2 for equality you must test each member of each structure for equality, as in the statement

```
if ( s1.a == s2.a && s1.b == s2.b && s1.c == s2.c )
```

2. A structure pointer can be used to access a structure variable. The structure pointer operator is -> and not - > (no spaces are allowed between - and >).
3. Beware of the differences between typedef and #define. As an example, consider the following:

```
typedef int * INT_PTR ;
INT_PTR p1, p2 ;
```

In the definition above both p1 and p2 are pointers to integers.
Now consider the following:

```
#define INT_PTR int *
INT_PTR p1, p2 ;
```

The preprocessor will translate the definition of p1 and p2 above as:

```
int *p1, p2 ;      (not int *p1, *p2 ; )
```

Here p1 is a pointer to an integer, as before. However, p2 is an integer and not a pointer to an integer. The #define preprocessor directive simply replaces INT_PTR with int *, but typedef allows you to give a new name to an existing data type.
4. The members of a structure do not necessarily occupy adjacent memory locations. Consider the following structure:

```
struct
{
  char m1 ;
  int m2 ;
  char m3 ;
} s ;
```

1. 即使两个结构体变量是用相同的结构体模板来定义的，也不可以在if语句中直接比较这两个结构体变量。

若要比较两个结构体变量相等与否，那么必须对两个变量的每个成员都单独进行比较。

2. 访问结构体指针指向的结构体变量需要使用指向运算符->，而不是- >（-和>之间不允许有空格）。
3. 注意typedef和#define之间的区别。

使用#define编译预处理指令的结果是，编译器简单地将INT_PTR替换为int *，而typedef只是给已有的数据类型定义一个别名。

4. 结构体变量的成员不一定占据相邻的内存空间。

在某些计算机中，结构体变量s的成员在内存中是按如下方式存储的。

也就是将结构体变量的所有成员存储在相邻的存储空间中。但是，在另一些

Some computers will store the members of the structure s as follows:

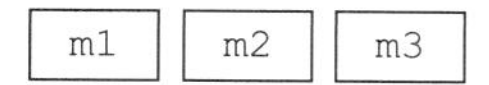

Here all the members of the structure are allocated adjacent memory locations. Some computers, however, require that all data items except character data start on an even address at a four-byte boundary. In this case the structure members could be allocated memory as follows:

In order to satisfy the alignment requirements of some computers, three bytes are skipped after `m1` so that the integer member `m2` starts on an even address at a four-byte boundary. The size of the structure variable is now one bigger than expected, and the members do not occupy adjacent memory locations. The compiler will take care of these details for you when you use the . and -> operators. However, do not try to access `m2`, for instance, by simply adding 1 to the address of `m1`. It may work on some computers but not on others.

You can check this out yourself by displaying the size of the variable s using the `sizeof()` operator.

计算机中，系统将不包括字符型在内的所有数据项都存储在以4字节为边界的偶数地址开始的存储空间中，此时结构体成员的内存空间分配情况如下。为了适应某些计算机系统的对齐要求，3个字节的空间被跳过去了，这样整型数据成员m2将存储在以4字节为边界的偶数地址开始的存储空间中，结构体变量占据的内存空间大小也就比预期的要多，结构体变量成员也就因此没有存储在相邻的存储空间中。当使用运算符.和->时，编译器会自动为用户注意这些细节问题，但是，注意不要试图使用类似于将m1的地址值加1的方式来访问m2，因为这样的方式在某些计算机系统中是适用的，而在另一些计算机系统中则是不适用的。

Quick syntax reference

	Syntax	Examples

Declaring a structure template	`struct structure_tag`
	`{`
	`  type variable`$_1$` ;`
	`  type variable`$_2$` ;`
	`  ...`
	`} ;`
		`struct date`
		`{`
		`  int day ;`
		`  int month ;`
		`  int year ;`
		`} ;`

Defining structure variables	`struct structure_tag variable`$_1$`,`
	`variable`$_2$`,`
	`... ;`
		`struct date dob ;`

Defining a pointer to a structure variable	`struct structure_tag *variable;`	`struct date *ptr ;`
Accessing structure members	`Dot operator` `-> operator`	`dob.day ;` `ptr-> dob.year ;`

Exercises

1. Write a structure template for each of the following:
   (a) a playing card, such as the five of diamonds or the three of spades
   (b) a stock record consisting of a stock number (integer), a stock description (20 characters), and a stock quantity (integer)
   (c) a library book record consisting of ISBN (13 characters), book title (30 characters), author (25 characters), and a price (`float`)
   (d) a customer record consisting of a customer number (`unsigned int`), a name (25 characters), an address (45 characters), and an outstanding balance (`double`)
   (e) a transaction record consisting of a transaction type (1 character), the date of the transaction (3 integers), and the amount of the transaction (`float`)
   (f) the time of day using the twelve-hour format, i.e. hours, minutes and seconds and either a.m. or p.m.
   (g) the longitude and latitude co-ordinates of a geographical position consisting of degrees (`int`), minutes (`int`), and direction (`char`) ('N', 'S', 'E' or 'W')
   (h) thirty teams in a league. For each team store the team name (20 characters) and the number of wins, draws and losses for both home and away games.
2. Given the following:

```
struct data
{
  int a ;
  float b ;
} ;
struct data d, *p = &d ;
```

   write statements that assign `1` to `a` and `2.3` to `b`, using
   (a) the `.` operator
   (b) the `->` operator
   (c) the `*` operator.
3. Given the following definitions:

```
struct stock_record
{
  int no ;
  char description[21] ;
  float price ;
  int qty ;
} ;

struct stock_record stock_item ;
```

   write statements to:
   (a) assign a value to each member of `stock_item`
   (b) input a value to each member of `stock_item`
   (c) display the value of each member of `stock_item`.
4. Create an enumerated data type for each of the following:
   (a) the days of the week: Monday, Tuesday, Wednesday, and so on

(b) the months of the year

(c) monetary denominations

(d) the suits in a pack of cards

(e) the points on a compass.

5. Given the following structure template

```
struct identification
{
  char name[21] ;
  float height ;
  float weight ;
  char hair_colour[11] ;
  char eye_colour[11] ;
} ;
```

and the following definition of the variable `id`,

```
struct identification id ;
```

and the following function calls,

```
input_data( &id ) ;
display_data( const id ) ;
```

write the functions `void input_data( struct identification* )` and
`void display_data( const struct identification )`.

6. In question 5, what changes would you make if the second function call was

```
display_data ( &id ) ;
```

Why would this function call be more efficient?

7. Given the following definitions

```
struct time_rec
{
  int hours ;
  int mins ;
  int secs ;
} ;
struct time_rec current_time ;
```

write a program containing functions to do the following:

(a) enter values for `current_time`

(b) add one second to `current_time`

(c) display the new values in `current_time`.

8. Given the following:

```
struct date_rec
{
  int day ;
  int month ;
  int year ;
} ;
```

```
struct date_rec current_date ;
```

write a program containing functions to do the following:

(a) enter values for `current_date`

(b) add one day to `current_date`

(c) display the values in `current_date`.

For the purposes of this exercise, assume that there are thirty days in each month.

9. Modify the program in exercise 8 to take into account the actual number of days in each month. Also, allow for leap years.

10. Using the structure template `date_rec` in exercise 8, write a function that accepts two dates and returns `0` if the two dates are equal, `1` if the first date is later than the second date, and `-1` if the first date is earlier than the second date.

Chapter Thirteen
File Input and Output
第 13 章 文件的输入和输出

Up to this point, the input-output statements used were those that read data from the keyboard and displayed data on the screen. When a program reads data from the keyboard, it stores the data in the computer's memory. When the program terminates, the data is lost and must be re-entered every time the program is run. This chapter covers standard file input and output. Unlike data that is kept in the computer's memory, files use external storage devices, such as hard disks and USB keys to store data. These are permanent storage devices that allow data to be stored after the program terminates.

到目前为止，我们使用的输入语句都是从键盘输入的，输出语句都是向显示器屏幕输出的。当程序从键盘输入数据后，数据存储在计算机的内存中，当程序结束时，这些数据就丢失了。因此，每次运行程序时，都要重新输入数据。本章主要讲述的内容是标准文件输入和输出，与在计算机内存中保存数据不同的是，文件使用外部存储设备来存储数据，这些设备包括硬盘、闪存盘等，它们都是永久性的存储设备，即在这些设备上存储的数据在程序结束后不会丢失。

13.1 Binary and ASCII (text) files [二进制文件和ASCII（文本）文件]

There are two types of files in C: *text* (or *ASCII*) files and *binary* files. The difference between the two types of files is in the way they store numeric-type (`int`, `float`, etc) data. In binary files, numeric data is stored in binary format, while in text files numeric data is stored as ASCII characters. For example, if you had

在C语言中，文件有两种类型：**文本文件**（或**ASCII文件**）和**二进制文件**。两种文件的区别在于，它们存储数值类型（整型、浮点型等）数据的方式不同。在二进制文件中，数值类型数据是按二进制形式存储的；而在文本文件中，数值类型数据是按照ASCII字符形式存储的。

```
short int n = 123 ;
```

The variable `n` occupies two bytes of memory. (See program P2D.)

这里，变量n在内存中占两个字节。

The value of `n` in a text file, as shown in Figure 13.1, requires three bytes of memory.

若将n存储到文本文件中，则需要3个字节的存储空间。

Character:	'1'	'2'	'3'
ASCII value in decimal:	49	50	51
ASCII value in binary:	00110001	00110010	00110011

Figure 13.1 text file storage for 123

Each digit of the number requires one byte of storage in an ASCII file. In a binary file, the digits of a number do not occupy individual storage locations. Instead the number is

在ASCII文件中，数值的每一位数字都要占据一个字节的存储空间。
而在二进制文件中，数值的每一位并不占据一个独立的存储空间，数值是当成一个整体来存储的。
如果将n值增大到1234，那么在二进制文件中，变量n所需的存储空间将仍然与其值为123时相同。但是，在ASCII文件

stored in its entirety as a binary number. So the variable n, with a value of 123, will be stored as shown in Figure 13.2.

00000000	01111011

Figure 13.2 binary file storage for 123

If, for example, you increase the value of n to 1234, the binary file will still require the same storage as for the number 123. However, an ASCII file would require one more byte to store the extra digit.

Which type of file should you use? There are advantages to both types of files. ASCII files are easily displayed on screen using simple system utilities, like the Windows type command or the Linux cat command, and so the files created by a program can easily be checked. Another advantage of text files is that they are easily read by other programs, such as text editors, word-processors, spreadsheets, database systems, and so on.

A disadvantage of using text files is that numeric data must be converted from ASCII to binary when it is read from a text file and converted from binary to ASCII when it is written to a text file. A binary file, on the other hand, does not require any conversions and may take up less space on the disk than a text file containing the same numeric data. However, although more efficient than text files, binary files cannot be readily read by other non-C programs.

中，存储变量n所需的存储空间需要额外再增加1个字节来存储多出来的一位数字4。

那么究竟要选择哪种文件类型呢？两种文件都各有其优势。对于文本文件而言，使用简单的系统实用工具就可以很容易地将其文件内容显示到屏幕上，这类工具包括Windows的type命令，或者Linux的cat命令等，因此可以很方便地检查程序创建的这类文件。文本文件的另一个优点是，它们可以很方便地被其他程序读取，包括文本编辑器、Word文档处理器、电子制表软件、数据库系统等。

13.2 Opening and closing files（文件的打开和关闭）

Before you can access a file for input or output, you must first open that file. To open a file in C, you use the standard library function **fopen( )**.

在对文件进行输入、输出操作之前，必须先打开这个文件。在C语言中，打开文件需要使用标准库函数fopen( )。

Program Example P13A

```
1  /* Program Example P13A
2     Demonstration of opening and closing a file.*/
3  #include <stdio.h>
4  int main()
5  {
6    FILE *fp ;      /* fp is a pointer to a file. */
7
8    /* Open the file named file.txt for reading. */
9    fp = fopen( "file.txt", "r" ) ;
10   /* If unable to open the file, then fp is NULL. */
11   if ( fp == NULL )
12   {
13     puts( "Error in opening file.txt" ) ;
14   }
15   else
16   {
17     puts( "file.txt is successfully opened" ) ;
```

```
18      fclose( fp ) ;
19    }
20    return 0 ;
21 }
```

Line 6 defines the variable fp as a pointer to a file. When the file is opened, the file pointer fp will point to the data at the start of the file. As data is read or written to the file, fp will automatically point to the next byte of storage in the file. Note that FILE is in uppercase. FILE is a data structure that is already defined in stdio.h.

程序第6行定义变量fp为指向文件的指针。打开文件时，文件指针fp将指向文件的起始位置，随着数据的读入或写入，fp将自动指向文件中的下一个字节。注意，FILE中的字母都是大写的，FILE是一种已在stdio.h中定义的数据结构。

Line 9 calls the function fopen() to open the file named file.txt. The fopen() function takes two arguments: the name of the file to open and a file *mode*. Table 13.1 contains the possible values for the file mode.

程序第9行调用函数fopen( )来打开一个名为file.txt的文件。函数fopen( )有两个实参：要打开的文件名（用双引号括起来），文件打开**模式**（也是用双引号括起来的）。

Table 13.1 file modes

Mode	Action
"r"	Open a text file for reading only. The file must already exist, or else the file pointer is NULL.
"w"	Open a text file for writing only. The file is created if it does not already exist. An existing file is replaced.
"a"	Open a text file for writing. Data is appended to the end of the file, or a new file is created if the file does not already exist.
"r+"	Open a text file for both reading and writing. The file must already exist, or else the file pointer is NULL.
"w+"	Open a text file for both writing and reading. The file is created if it does not already exist. An existing file is replaced.
"a+"	Open a text file for both reading and appending. Data is written to the end of the file, or a new file is created if the file does not already exist.

In addition, the letter b may be added to the modes listed above to indicate that a file is to be opened in binary mode. For example, the mode "rb" opens a binary file for reading, and the mode "a+b"(or "ab+") opens a binary file for reading and appending.

此外，字母b也可以添加到表中所列的模式中，表示要打开的文件是二进制文件，例如，模式rb表示以只读的方式打开二进制文件，模式"a+b"（或"ab+"）表示以读写的方式打开二进制文件。

If the file is successfully opened in the mode specified, fp points to the start of the file. However, if the file cannot be opened in the specified mode, fopen() returns NULL to the file pointer fp. Like FILE, NULL is defined in stdio.h. In program P13A, line 9 attempts to open the file file.txt for reading. Line 11 tests the value of fp. If the file pointer fp is NULL, the file cannot be opened for reading (perhaps the file does not exist), and an error message is displayed in line 13. If the file is successfully opened, a message is displayed in line 17, and the file is closed with the standard library function fclose() on line 18.

如果文件按指定的方式打开成功，那么fp将指向文件的起始位置。但是，如果文件不能按指定的方式打开，那么fopen( )将返回NULL给文件指针fp。与FILE一样，NULL也在stdio.h中定义。

The general format of fopen() is:

```
file_pointer = fopen( "filename", "mode" ) ;
```

The general format of fclose() is:

```
fclose( file_pointer ) ;
```

where file_pointer is any variable of the type FILE*.

Run the program P13A. If you have a file named file.txt, the file will be opened; otherwise you will get the error message from line 13.

It is quite common in C programming to combine the opening of a file and the testing of the file pointer. This is demonstrated in the next program.

在C程序中，在打开文件的同时检测文件指针是很常见的一种用法。

Program Example P13B

```
1  /* Program Example P13B
2     Opening and closing a file in a C program.    */
3  #include <stdio.h>
4  int main()
5  {
6    FILE *fp ; /* fp is a pointer to a file.      */
7
8    /* Open the file named file.txt for reading. */
9    if ( (fp = fopen("file.txt", "r")) == NULL )
10   {
11     puts( "Error in opening file.txt" ) ;
12   }
13   else
14   {
15     puts( "file.txt is successfully opened" ) ;
16     fclose( fp ) ;
17   }
18   return 0 ;
19 }
```

Line 9 of this program replaces lines 9 and 11 of program P13A. Note that the three sets of parentheses are necessary, because == has a higher precedence than =.

注意，由于关系运算符==的优先级比赋值运算符=的优先级高，所以这里的三对圆括号是必须的。

13.3 Reading a character from a file using fgetc() [使用函数fgetc( )从文件中读字符]

The standard library function **fgetc( )** reads a single character from a file. It is the file version of getchar(). The general format of fgetc() is:

标准库函数fgetc( )用于从文件中读取单个字符，它是函数getchar( )的文件操作版。

```
char_in = fgetc( file_pointer ) ;
```

This function reads a single character from the file opened with the file pointer file_pointer. The function places the character read from the file in the variable char_in. The value of char_in becomes **EOF** if there is an error or the end of file is reached. EOF is defined as -1 in the included header file stdio.h.

The next program demonstrates the function fgetc() by reading and displaying the entire contents of a text file file.txt.

该函数首先从已打开的由文件指针file_pointer指向的文件中读取单个字符，然后将从文件中读取的这个字符存放到变量char_in中。如果出现错误或者已经读到文件末尾，则该函数返回EOF，EOF在包含的头文件stdio.h中定义，其值为-1。

Program Example P13C

```
1  /* Program Example P13C
2     Demonstration of the function fgetc() by reading
3     the contents of a file character by character.  */
4  #include <stdio.h>
5  int main()
6  {
7    FILE *fp ;
8    char char_in ;
9
10   if ( (fp = fopen("file.txt", "r")) == NULL )
11   {
12     puts( "Error in opening file.txt" ) ;
13   }
14   else
15   {
16     while( (char_in = fgetc(fp)) != EOF )
17       putchar( char_in ) ;
18     fclose( fp ) ;
19   }
20   return 0 ;
21 }
```

Line 10 opens the file `file.txt`, and line 12 displays an error message if the opening is unsuccessful. Lines 16 and 17 form a `while` loop. The `fgetc()` in line 16 reads the next character from the file into `char_in`, and line 17 displays `char_in`. The loop continues until the end of the file is reached, in which case `char_in` will have the value `EOF`.

程序第10行语句打开文件file.txt，如果文件打开失败，则程序第12行显示一个出错信息。第16行和第17行构成一个while循环。第16行的函数fgetc( )从文件中读取下一个字符，存到变量char_in中，第17行显示变量char_in的值。重复执行该循环，直到读到文件末尾为止，此时变量char_in的值为EOF。

Strictly speaking, `fgetc()` returns `EOF` when either the end of the file is reached or an error has occurred. So the program above would appear to work fine even if there was an error in the file!

严格地讲，无论是读到文件末尾，还是出现文件读取错误，函数fgetc( )都将返回EOF。所以即使出现文件错误，上面的程序也会看似正常工作。

The **ferror( )** function is used to check whether a file error has occurred. `ferror()` returns a non-zero value if an error has occurred; otherwise it returns a 0. You could use this function in program P13C by placing the following after line 17:

```
if ( ferror(fp) )   /* Error on file? */
   puts( "Error on file" ) ;
```

函数ferror( )用来检测是否出现文件错误，如果出现错误，则函数返回一个非零值，否则，函数返回零值。

The **feof( )** function could also be used in program P13C. `feof()` allows you to check whether the end of the file has actually been reached. `feof()` returns a 0 if the end of the file has not been reached, otherwise it returns a non-zero value. You could use `feof()` in program P13C by placing the following after line 17:

```
if ( !feof(fp) )    /* End of file not reached? */
   puts( "Error on file" ) ;
```

13.4 Writing a character to a file using `fputc()` [使用函数fputc( )向文件中写字符]

The standard library function **fputc( )** writes a single character to a file; it is the file version `putchar()`. The general format of `fputc()` is:

标准库函数fputc( )用于向文件中写入单个字符，它是函数putchar( )的文件操作版。

```
fputc( char_out, file_pointer ) ;
```

`fputc()` writes the character `char_out` to the file opened for writing with the file pointer `file_pointer`.

函数fputc( )向已打开的由文件指针file_pointer指向的文件中写入字符char_out。

Program Example P13D

```
1  /* Program Example P13D
2     This program copies a file character by character using.
3     fgetc() and fputc().                                    */
4  #include <stdio.h>
5  int main()
6  {
7    FILE *fp_in, *fp_out ;
8    char char_in ;
9
10   if ( (fp_in=fopen("file.txt", "r")) == NULL )
11   {
12     puts( "Error in opening file.txt" ) ;
13   }
14   else if ( (fp_out = fopen("new.txt", "w"))!= NULL )
15   {
16     /* Proceed with the copying. */
17     while ( (char_in = fgetc(fp_in)) != EOF )
18     {
19       fputc( char_in, fp_out ) ;
20     }
21     fclose( fp_in ) ;
22     fclose( fp_out ) ;
23     puts( "Copying completed." ) ;
24   }
25   else
26   {
27     puts( "Error in opening new.txt" ) ;
28   }
29   return 0 ;
30 }
```

This program first opens the file `file.txt` for reading on line 10. If no error occurs, the file `new.txt` is opened for writing on line 14. If this file is successfully opened, the `while` loop between lines 17 to 20 reads a character from `file.txt` and writes this character to `new.txt`. The `while` loop continues until `char_in` has the value EOF.

13.5 Reading a string of characters from a file using fgets() [使用函数 fgets( ) 从文件中读字符串]

The standard library function **fgets( )** reads a string of characters from a file. It is the file equivalent of the `gets()` function, which reads a string of characters from the keyboard. The general format of `fgets()` is:

标准库函数fgets( )用于从文件中读取一个字符串。它是函数gets( )的文件操作版，函数gets( )的作用是从键盘读取一个字符串。

```
fgets( string_pointer, max_characters, file_pointer ) ;
```

The first argument, `string_pointer`, is a pointer to a character array used to store the character string. The second argument, `max_characters`, is the maximum number of characters in the string (including the `'\0'`) and the third argument, `file_pointer`, is a file pointer for a file opened for reading. The function reads characters from the file into an array until one of the following three conditions occur:

函数的第一个实参string_ pointer是一个指向字符数组的指针，该指针指向的字符数组用于存储从文件中读取的字符串，第二个实参max_characters 是指从文件读取的字符串的最大长度(包括字符串结束符'\0' 在内)，第三个参数file_pointer 是一个文件指针，它指向以读方式打开的文件。该函数从文件读取字符串，然后将其写入字符数组中，当遇到下列情况之一时结束操作：

a newline character (`'\n'`) is read
or the end of the file is reached
or (`max_characters-1`) characters are read.

(1) 读到换行符'\n'
(2) 读到文件末尾
(3) 已经读取了max_characters –1个字符

In all cases, the null character (`'\0'`) is placed in the array after the last character read. Unlike `gets()`, `fgets()` includes the newline character (`'\n'`) in the string.
The `fgets()` function returns the value of its first argument, `string_pointer`, unless an error or end of file occurs, in which case `NULL` is returned. Like `gets()`, `fgets()` stops reading when the newline character (`'\n'`) is read. This makes `fgets()` convenient for reading an entire line from a text file.

无论在上述哪种情况下，函数都会在读取的最后一个字符后面加上字符串结束符('\0')。与函数gets( )不同的是，函数fgets( )会将换行符('\n')也读到字符串中。
函数fgets( )返回它的第一个实参指针string_pointer 的值，当函数读到文件末尾或者出现读取错误时，函数返回空指针NULL。与gets( )类似，函数fgets( )在遇到换行符时结束。函数fgets( )的这个特点使得它很适合从文本文件中读取一整行的字符。

Program Example P13E

```
1   /* Program Example P13E
2      This program reads one line at a time from a file.
3      It assumes that no line is more than 80 characters long. */
4   #include <stdio.h>
5   #define MAX_CHARACTERS 81
6   int main()
7   {
8     FILE* fp_in ;
9     char one_line[ MAX_CHARACTERS ] ;/* Array used to store line. */
10
11    /* Open the file for reading */
```

```
12   if ( (fp_in=fopen("file.txt", "r")) == NULL )
13   {
14     puts( "Error in opening file.txt" ) ;
15   }
16   else
17   {
18     /* Read at most MAX_CHARACTERS-1 characters from the file
19        or until a new line character (\n) is read
20        or the end of file occurs. */
21     while( fgets( one_line, MAX_CHARACTERS, fp_in )!=NULL )
22       printf( "%s", one_line ) ;
23     fclose( fp_in ) ;
24   }
25   return 0 ;
26 }
```

The file `file.txt` is opened for reading on line 12. The loop in lines 21 and 22 displays each line of the file. The loop continues until `fgets()` returns `NULL` at the end of the file.

13.6 Writing a string of characters to a file using fputs() [使用函数fputs( )向文件中写字符串]

The function **fputs( )** writes a string of characters to a file opened for writing. The general format of `fputs()` is:

标准库函数fputs( )用于向以只写方式打开的文件中写入一个字符串。

```
fputs( string_pointer, file_pointer ) ;
```

where `string_pointer` is a pointer to a character string and `file_pointer` is a file pointer for a file opened for writing.

The null character (`'\0'`) is not written to the file, and, unlike its keyboard equivalent, `puts()`, `fputs()` does not add the newline character (`'\n'`) to the end of the string.

函数fputs( )不会在文件中写入空字符('\0')，并且与其相应的键盘操作函数puts( )不同的是，函数fputs( )不会在字符串的末尾加上换行符。

The next program uses both `fgets()` and `fputs()` to copy the contents of a file line by line.

Program Example P13F

```
1  /* Program Example P13F
2     This program reads and writes one line at a time using
3     fgets() and fputs(), assuming a line is not more than
4     80 characters long.                                    */
5  #include <stdio.h>
6  #define MAX_CHARACTERS 81
7  int main()
8  {
9    char one_line[ MAX_CHARACTERS ] ;
10   FILE *fp_in, *fp_out ;
11
12   if ( (fp_in = fopen("file.txt", "r")) == NULL )
```

```
13     {
14       puts( "Error in opening file.txt" ) ;
15     }
16     else if ( (fp_out = fopen("new.txt", "w")) != NULL )
17     {
18       /* Proceed with the copying. */
19       while( fgets(one_line, MAX_CHARACTERS, fp_in) != NULL )
20         fputs( one_line, fp_out ) ;
21       fclose( fp_in ) ;
22       fclose( fp_out ) ;
23       puts( "Copying completed." ) ;
24     }
25     else
26     {
27       puts( "Error in opening new.txt" ) ;
28     }
29     return 0 ;
30   }
```

This program is similar to P13E, except that instead of displaying the line on the screen, the `fputs()` function on line 20 writes the line to the file `new.txt`.

13.7 Formatted input-output to a file using fscanf() and fprintf() [使用函数fscanf( )和fprintf( )进行文件的格式化读写]

The **fscanf( )** and **fprintf( )** functions work just like the `scanf()` and `printf()` functions, except that they read and write data to a file. The general format of `fprintf()` and `fscanf()` is:

函数fscanf( )和fprintf( )与函数scanf( )和printf( )的功能类似，不同的是它们是对文件进行读写。

```
fprintf( file_pointer, format_string, variable₁, variable₂,... ) ;
fscanf ( file_pointer, format_string, variable₁, variable₂,... ) ;
```

The first argument, `file_pointer`, is a file pointer. The second argument, `format_string`, describes the format used when reading or writing the variables `variable`$_1$, `variable`$_2$, etc. The format string contains the same formatting commands used by `printf()` and `scanf()`.

`fprintf()`, like `printf()`, returns the number of data items successfully written.

`fscanf()`, like `scanf()`, returns the number of items successfully read, or `EOF` (`-1`) when the end of the file or an error occurs.

Both `fscanf()` and `printf()` are used in the next program to copy the contents of a file `sales.dat` to a file `newsales.dat`.

第一个实参file_pointer是一个文件指针，第二个实参format_string为格式字符串，用于指定variable1、variable2等变量的读写格式，该格式字符串中包含的格式指令与printf( )和scanf( )包含的格式指令相同。

与函数printf( )类似，函数fprintf( )的返回值为成功写数据的个数。

与函数scanf( )类似，函数fscanf( )的返回值为成功读数据的个数，若函数读到文件末尾或者出现读取错误，则函数返回EOF（值为-1）。

Program Example P13G

```
1  /* Program Example P13G
2     Demonstration of fscanf() and fprintf().
```

```
3     This program reads a sales file, displays the data on
4     screen, and displays the total sales value.
5     The sales data is also copied to a new file.       */
6  #include <stdio.h>
7  #define NAME_SIZE 31
8  int main()
9  {
10   FILE *fp_in, *fp_out ;
11   char name[ NAME_SIZE ] ;  /*  Product name.   */
12   int quantity ;            /*  Quantity sold. */
13   float value ;             /*  Sales value.   */
14   float total_value = 0.0 ;
15
16   if ( (fp_in = fopen("sales.dat", "r")) == NULL )
17   {
18     puts( "Error in opening file sales.dat ") ;
19   }
20   else if ( (fp_out = fopen("newsales.dat", "w")) != NULL )
21   {
22     /* Read input file until EOF is encountered. */
23     while ( (fscanf(fp_in,"%s%d%f",name,&quantity,&value))!=EOF )
24     {
25       fprintf( fp_out,"%s %d %6.2f\n", name, quantity, value ) ;
26       printf( "%s %d %6.2f\n", name, quantity, value ) ;
27       total_value += value ;
28     }
29     printf( "\nTotal sales value = %7.2f\n", total_value ) ;
30     fclose( fp_in ) ;
31     fclose( fp_out ) ;
32   }
33   return 0 ;
34 }
```

Line 23 uses `fscanf()` to read the product name, quantity sold, and sales value from the file `sales.dat`. Line 25 uses `fprintf()` to write the data to the new file `newsales.dat`, and line 26 displays the data on the screen. The `fscanf()` function in line 23 reads the sales quantity and value in ASCII and converts them to binary. The `fprintf()` function in line 25 does the opposite, by converting numeric binary data to their ASCII equivalents.

The `while` loop in lines 23 to 28 stops when `fscanf()` returns `EOF` at the end of the file. At the end of the file, line 29 displays the total sales value. The files are closed in lines 30 and 31 and the program stops.

程序第23行使用函数fscanf( )从文件sales.dat中读取产品的名称、销售数量、销售金额。第25行将这些数据写入一个新的文件newsales.dat中。第26行将这些数据显示到屏幕上。第23行的函数fscanf( )读取销售数量和销售金额时，将其从ASCII码形式转换为二进制数值形式，第25行的函数fprintf( )则相反，它将二进制数值形式的数据转换为等价的ASCII码形式。

程序第23行~第28行的while循环，在函数fscanf( )读到文件末尾、函数返回EOF时结束。文件数据读完以后，程序的第29行语句在屏幕上显示总的销售金额。第30行和第31行

13.8　The standard files（标准文件）

When a C program starts, the operating system automatically opens three standard files as shown in Table 13.2 below and associates a file pointer with each file.

Table 13.2 standard files

File name	File pointer	Assigned to
Standard input	stdin	keyboard
Standard output	stdout	screen
Standard error	stderr	screen

关闭文件sales.dat和newsales.dat，然后程序结束。

当C程序运行时，操作系统会自动打开如表13.2所示的3个标准文件，并将一个文件指针与每个文件进行关联。

Thus,

```
printf( "%s\n", "Hello" ) ;
```

is equivalent to

```
fprintf( stdout, "%s\n", "Hello" ) ;
```

and

```
scanf( "%d", &n ) ;
```

is equivalent to

```
fscanf( stdin, "%d", &n ) ;
```

Just as the files associated with these pointers are opened automatically when a program starts, they are closed automatically when the program finishes.

当程序开始运行时，与这些文件指针相关联的文件将会被自动打开。同样，当程序结束时，则自动关闭这些文件。

13.9 Block input-output using fread() and fwrite() [使用函数fread( )和fwrite( )进行块读写]

Block (or *binary*) *I-O* does not restrict you to reading a character, a word or a line at a time. With block I-O you specify the size of the block of memory you want to read or write. The block size can be as low as one byte or as large as an entire file.

Instead of translating numeric data into ASCII, the **fwrite( )** function simply copies the data in a specified block of memory to a file without doing any translation. Similarly, the **fread( )** function will read a block of data from a file and copy it to a memory location without doing any translation from ASCII to binary. The contents of the file are, in fact, a binary image of a block of memory.

The general format of `fread()` and `fwrite()` is:

块（或二进制数据）读写，使我们不再局限于一次只读取一个字符、一个单词或是一行字符串。块读写允许用户自己指定想要读写的内存块的大小，这个数据块的大小最小可以是1 个字节，最大可以是整个文件。

函数fwrite( )只是简单地将指定内存块中的数据复制到文件中，它不做任何转换工作，也就是说它不会将数值数据转换为ASCII数据。同样，函数fread( )也只是直接从文件中读取一个数据块，然后将其复制到内存单元中，它不做从ASCII数据转换为二进制数据的工作。事实上，文件的内容就是内存块的二进制映像。

```
num_read = fread( memory_pointer, size, num, file_ptr_in ) ;
num_written = fwrite( memory_pointer, size, num, file_ptr_out ) ;
```

The first argument, `memory_pointer`, is a pointer to a memory location where the data items are stored. This will normally be an array name or a pointer to a structure variable. The second argument, `size`, is the number of

第一个实参memory_pointer是指向存储数据项的内存空间的指针，它通常是一个数组名或是一个指向结构体变量的指针。第2个实参size是单个数据项所占的字节

bytes in each data item. Usually the `sizeof` operator is used to find the size of a data item.

数，通常使用运算符sizeof来计算单个数据项所占内存空间的大小。

The third argument, `num`, specifies the number of data items that are to be read or written.

第3个实参num指定要读写的数据项的个数。

The fourth argument is a file pointer for an open file.

第4个实参是指向已打开文件的指针。

Both functions return an integer value, which is the actual number of data items read or written. If the returned value is less than `num`, an error has occurred or, in the case of `fread()`, the end of the file has been reached.

上述两个函数都返回一个整型值，该值为实际读写的数据项的个数。如果函数的返回值小于num，那么可能是读写过程中出现了错误，或者是在使用fread( )函数时读到了文件的末尾。

The next program writes an array of integers to a file using `fwrite()` and reads the array back again using `fread()`.

Program Example P13H

```
1   /* Program Example P13H
2      Demonstration of block I-O routines fread() and fwrite().
3      Program writes an array of ten integers to a file and
4      reads them back again.                                      */
5   #include <stdio.h>
6   #include <stdlib.h>
7   #define SIZE 10
8   int main()
9   {
10    FILE *fp ;
11    int count, i ;
12    int array[SIZE] = { 0, 1, 2, 3, 4, 5, 6, 7, 8, 9 } ;
13
14    /* Open the output file. */
15    if ( (fp = fopen("array.dat", "wb")) == NULL )
16    {
17      puts( "Error on opening output file array.dat" ) ;
18      return EXIT_FAILURE ;
19    }
20
21    /* Write the array to the file. */
22    count = fwrite( array, sizeof(int), SIZE, fp ) ;
23
24    if ( count != SIZE )
25    {
26      puts( "Error on writing array.dat" ) ;
27      return EXIT_FAILURE ;
28    }
29    fclose( fp ) ;
30
31    /* Open the file for input. */
32    if ( (fp = fopen("array.dat", "rb")) == NULL )
33    {
34      puts( "Error on opening input file array.dat" ) ;
35      return EXIT_FAILURE ;
```

```
36    }
37
38    /* Read the array from the file. */
39    count = fread( array, sizeof(int), SIZE, fp ) ;
40    if ( count != SIZE )
41    {
42      puts( "Error on reading array.dat" ) ;
43      return EXIT_FAILURE ;
44    }
45    fclose( fp ) ;
46
47    puts( "The array values are:" ) ;
48    for ( i = 0 ; i < SIZE ; i++)
49      printf( "%d ", array [i] ) ;
50    putchar ( '\n' ) ;
51    return 0 ;
52 }
```

The output from this program is:

```
The array values are:
0 1 2 3 4 5 6 7 8 9
```

程序第15行以二进制只写方式打开文件array.dat，如果文件打开不成功，则显示一个出错信息，在第18行用return语句结束程序的执行。返回给操作系统的状态码EXIT_FAILURE表示程序运行失败。

Line 15 opens the file `array.dat` for writing in binary mode. If the file cannot be opened, an error message is displayed, and the program is terminated with the `return` statement on line 18. The status code, `EXIT_FAILURE`, is passed to the operating system to indicate failure.

第22行将数组元素值写到文件array.dat 中。第24行检查是否所有数组元素都已写入文件中。如果数组元素并未全部成功写入文件中，那么在第27行用return语句结束程序的执行，否则在第29行关闭该文件，然后在第32行以二进制只读方式再次打开该文件。

Line 22 writes the array to the file `array.dat`. Line 24 checks that all the array elements have been written to the file. If all the array elements have have not been written to the file, the program exits with the `return` statement on line 27; otherwise line 29 closes the file, and line 32 opens it again for binary reading.

第39行将数组元素值重新读回到内存中，第48行和第49行显示数组的每个元素的值。

Line 39 reads the array back into memory, which is then displayed by lines 48 and 49.

13.10 Rewinding a file using `rewind()` [使用函数rewind( )对文件重定位]

函数rewind( )将当前的文件位置重定位到文件的开始处。

The **rewind( )** function resets the current file position to the start of the file. The general format of `rewind()` is:

```
rewind( file_pointer ) ;
```

where `file_pointer` is a file pointer to an open file.

The next program is a modification of program P13H. Instead of closing the file after writing is completed (line 29) and opening the file again for reading (line 32), `rewind()` is used.

Program Example P13I

```
1  /* Program Example P13I
2     Rewrite of program P13H using the rewind() function.
3     Program writes an array of ten integers to a file and
4     reads them back again.                                  */
5  #include <stdio.h>
6  #include <stdlib.h>
7  #define SIZE 10
8  int main()
9  {
10   FILE *fp ;
11   int count, i ;
12   int array[SIZE]={ 0,1,2,3,4,5,6,7,8,9 } ; /* Array of integers. */
13
14   /* Open the output file. */
15   if ( (fp=fopen("array.dat", "w+b")) == NULL )
16   {
17     puts( "Error on opening output file array.dat" ) ;
18     return EXIT_FAILURE ;
19   }
20
21   /* Write the array to the file. */
22   count = fwrite( array, sizeof(int), SIZE, fp ) ;
23   if ( count != SIZE )
24   {
25     puts( "Error on writing array.dat" ) ;
26     return EXIT_FAILURE ;
27   }
28
29   rewind( fp ) ;
30
31   /* Read the array from the file. */
32   count = fread( array, sizeof(int), SIZE, fp ) ;
33   if ( count != SIZE )
34   {
35     puts( "Error on reading array.dat" ) ;
36     return EXIT_FAILURE ;
37   }
38   fclose( fp ) ;
39
40   puts( "The array values are:" ) ;
41   for ( i = 0 ; i < SIZE ; i++ )
42     printf( "%d\n", array [i] ) ;
43   putchar( '\n' ) ;
44
45   return EXIT_SUCCESS ;
46 }
```

In this program, the file `array.dat` is opened for binary reading and writing (mode `"w+b"`) on line 15. Instead of closing the file and reopening

在程序第15行，以二进制读写方式打开文件array.dat。在写操作结束后，没有关闭文件再重新打开它，而是用第29行语句将文件的当前位置重定位到文件的开始处。然后在

it, line 29 rewinds the file to the start. The function `fread()` is then used on line 32 to read the data from the file back into memory.
The output from this program is the same as program P13H.

第32行使用函数fread( )从文件中将数据读回到内存中。

13.11 Random access of files using `fseek()` [使用函数fseek( )随机访问文件]

The previous programs all perform *serial file processing*. With serial file processing, data items are read or written one after the other. For example, if you wanted to read the fifth data item in a file, with serial processing you must read the previous four data items first. With *random* or *direct access* you can move around in a file, reading and writing at any position in the file.
The **fseek( )** function allows you to move to any position in a file, from where you can then read or write data. The general format of `fseek()` is:

```
fseek( file_pointer, offset, origin )
```

The first argument is a file pointer for a file opened for reading, writing, or both. The second argument, `offset`, is an integer and instructs `fseek()` where to move in the file. For example, the first byte in a file has an offset of 0 from the start of the file, the second byte has an offset of `1`, the third byte has an offset of `2`, and so on.
The third argument `origin` determines the point from where the offset is measured.
Table 13.3 shows the three possible values that `origin` can have.

前面的程序执行的都是对文件进行**串行处理**。在串行处理方式下，文件中的数据项是一个接一个地顺序读写的。例如，若要读取文件中的第5个数据项，按照这种串行方式，必须先读取前4个数据项才能读取第5个数据项。而使用**随机访问**或**直接访问**方式，我们可以在文件中随机定位，直接读取文件中任意位置上的数据项。
函数fseek( )允许移动文件位置指针到文件中的任意位置，从而实现在文件的任意位置上进行读写操作。
第1个实参是一个文件指针，指向以读、写或读写方式打开的文件。第2个实参offset是一个整型变量，指示移动到文件中的什么位置（即偏移量）。例如，文件中的第一个字节数据距离文件的开始位置的偏移量为0，第2个字节的偏移量为1，第3个字节的偏移量为2，以此类推。
第3个实参origin指定了偏移量计算的起始点。

Table 13.3 origin values and meanings

Origin	Meaning
`SEEK_SET` (defined as 0)	The offset is measured from the beginning of the file.
`SEEK_CUR` (defined as 1)	The offset is measured from the current position in the file.
`SEEK_END` (defined as 2)	The offset is measured from the end of the file.

Examples:

```
fseek( fp, 10, SEEK_SET ) ;  /* Offset is 10 from the start of
                                the file. So the next byte read
                                or written will be the 11th.   */
fseek( fp, 1, SEEK_CUR ) ;   /* Forward 1 byte from the current
                                position. So the next byte read
                                or written will be the 12th.   */
fseek( fp, -1, SEEK_CUR ) ;  /* Back 1 byte from the current
                                position. Now back at the 11th
                                byte in the file.              */
fseek( fp, -10, SEEK_END ) ; /* Offset is -10 from the end of
                                the file. The next byte read or
```

```
                                   written is 10 bytes before the
                                   end of the file.               */
  fseek( fp, 0, SEEK_END ) ;    /* End of file.                   */
  fseek( fp, 0, SEEK_SET ) ;    /* Start of file; equivalent to
                                   rewind( fp ).                  */
  fseek( fp, n, SEEK_SET ) ;    /* Go to the (n+1)th byte in the
                                   file, where n is an integer.   */
```

SEEK_SET, SEEK_CUR and SEEK_END are #defined as 0, 1 and 2, respectively, in the header file stdio.h. You can therefore use 0, 1 and 2 instead of SEEK_SET, SEEK_CUR and SEEK_END in the above examples. If fseek() is successful it returns the integer value 0; otherwise it returns a non-zero integer value.

The next program demonstrates the use of fseek() by displaying the ASCII value of any character in the text file file.txt.

SEEK_SET、SEEK_CUR和SEEK_END在头文件stdio.h中用#define分别将其定义为0、1和2。因此，也可以使用0、1和2分别代替SEEK_SET、SEEK_CUR和SEEK_END。如果函数fseek( )调用成功，则函数返回整数值0，否则它将会返回一个非零的整数值。

Program Example P13J

```
1 /* Program Example P13J
2    Demonstration of random access using fseek(). */
3 #include <stdio.h>
4 #include <stdlib.h>
5 int main()
6 {
7   FILE *fp ;
8   int file_pos, offset ;
9   int fseek_ret_value ;
10  char in_char ;
11
12  if ( (fp = fopen("file.txt", "r")) == NULL )
13  {
14    printf( "Cannot open file.txt" ) ;
15    return EXIT_FAILURE ;
16  }
17
18  printf( "\n\nDisplay the ASCII value of any character in "
19          "file.txt\n\n" ) ;
20  do
21  {
22    printf( "Enter the file position or 0 to end: ") ;
23    scanf( "%d", &file_pos ) ;
24
25    if ( file_pos > 0 )
26    {
27      /* First byte if offset 0, so reduce file_pos by 1. */
28      offset = file_pos - 1 ;
29      fseek_ret_value = fseek( fp, offset, SEEK_SET ) ;
30      if ( fseek_ret_value == 0 )
31      {
32        in_char = fgetc( fp ) ;
```

```
33         printf( "ASCII value at position %d is %d. ",
34                 file_pos, in_char ) ;
35         if ( in_char == ' ' )
36           printf( "This is the space character.\n" ) ;
37         else
38           printf( "This is the character %c\n", in_char ) ;
39       }
40       else
41         printf( "Invalid value\n" ) ;
42     }
43   } while ( file_pos > 0 ) ;
44
45   fclose( fp ) ;
46   return 0 ;
47 }
```

Line 23 of this program asks the user for a position in the file. Line 28 converts this position to an offset by subtracting 1. Line 29 uses `fseek()` to position the file for reading the next character, which is done in line 32. Line 30 checks the return value from `fseek()`. The return value `fseek_ret_value` is zero if the file location pointer is successfully moved; otherwise `fseek_ret_value` is non-zero.

程序的第23行请用户输入一个文件位置。第28行将其值减1得到文件位置的字节偏移量。第29行使用函数fseek( )定位到文件中下一个要读取的字符位置，第32行语句读取这个位置上的字符，第30行检查函数fseek( )的返回值，如果文件位置指针成功移动，则函数返回给变量fseek_ret_value的值为0，否则返回给变量fseek_ret_value一个非0值。

To demonstrate the use of `fseek()` and `fread()` on a binary file, consider a supplier file containing the following data:

```
CODE            NAME                    BALANCE
 1        Fruit Suppliers Ltd            100.00
 2        Jinsham Ltd                    210.50
 3        P. D. Baker                    155.50
etc.
 20       P. Last                         45.91
```

The above data is held in the binary file `supplier.dat`. Note that the supplier code corresponds to the position of the supplier record in the file, i.e. supplier code 1 is the first record, supplier code 2 is the second record and supplier code 20 is the twentieth record. This is illustrated in Figure 13.3 below.

1	Fruit Suppliers Ltd	100.00	2	Jinsham Ltd	210.50	3... etc

Figure 13.3 layout of file `supplier.dat`

The next program asks the user for a supplier number and displays the corresponding supplier record.

Program Example P13K

```
1 /* Program example P13K
2    To further demonstrate fseek() and fread().
3    This program reads a supplier code and displays the
4    corresponding supplier record.                          */
```

```
5  #include <stdio.h>
6  #include <stdlib.h>
7  FILE *open_file( const char *file, const char *mode ) ;
8  int get_input_code( void ) ;
9  void display_supplier( unsigned int supp_code, FILE *fp ) ;
10 #define MAX_RECS 20
11 int main()
12 {
13   FILE *fp ;
14   int in_code ;
15
16   /* Open the supplier file. */
17   fp = open_file( "supplier.dat", "rb" ) ;
18
19   /* Display supplier data for each code entered. Stop when
20      the supplier code entered is less than or equal to 0.     */
21   do
22   {
23     in_code = get_input_code() ; /* Get the supplier code.     */
24     /* If the code is in range, display the supplier data.     */
25      if ( in_code >0 && in_code <= MAX_RECS )
26        display_supplier( in_code, fp ) ;
27   } while ( in_code > 0 ) ; /* Loop ends when code is not > 0.*/
28   fclose( fp ) ;
29
30   return 0 ;
31 }
32
33 /* Function   : display_supplier()
34    Purpose    : This function displays the supplier record.
35    Parameters: Supplier code and file pointer.
36    Returns    : void.                                            */
37  void display_supplier( const unsigned int supp_code, FILE *fp )
38 {
39   struct supplier_rec
40   {
41    int code ;
42     char name[31] ;
43    float balance ;
44   } ;
45   struct supplier_rec supp ;
46   int offset ;
47
48   offset = ( supp_code - 1 ) * sizeof( struct supplier_rec ) ;
49   fseek( fp, offset, 0 ) ;
50
51   if ( (fread( &supp,sizeof(struct supplier_rec),1,fp )) != 1 )
52     printf( "\nError in reading file" ) ;
53   else
54     printf( "\nCode: %d\nName: %s\nBalance :%7.2f\n",
```

```
55              supp.code, supp.name, supp.balance ) ;
56 }
57
58 /* Function   : get_input_code()
59    Purpose    : Gets the supplier code from the keyboard.
60    Parameters: None.
61    Returns    : Supplier code.                            */
62 int get_input_code( void )
63 {
64   int supp_code ;
65
66   printf( "\nEnter a supplier code from 1 to %d or 0 to end ",
67           MAX_RECS ) ;
68   scanf( "%d", &supp_code ) ;
69   return ( supp_code ) ;
70 }
71
72 /* Function   : open_file()
73    Purpose    : Opens a file in the required mode.
74    Parameters: Filename and mode.
75    Returns    : File pointer. */
76 FILE * open_file( const char *file, const char *mode )
77 {
78   FILE *fp ;
79
80   if ( (fp=fopen(file, mode)) == NULL )
81   {
82     printf( "\n\nUnable to open file %s with %s\n",file,mode ) ;
83     exit( EXIT_FAILURE ) ;
84   }
85   return ( fp ) ;
86 }
```

On line 17, the function `open_file()` defined on lines 76 to 86 is used to open the supplier file.

In line 51, the number of bytes in each supplier record is got by using `sizeof(struct supplier_rec)`. Supplier 1 is the first record in the file and has an offset of 0. Supplier 2 is the second record in the file and has an offset of `1*sizeof(struct supplier_rec)`. Supplier 3 is the third record and has an offset of `2*sizeof(struct supplier_rec)`. Thus for any supplier code you can get the offset of the corresponding record by subtracting `1` from the supplier code and multiplying the result by `sizeof(struct supplier_rec)`. The calculation of the offset is done in line 48.

Once positioned at the correct place in the file, the `fread()` in line 51 reads the data into the structure variable `supp`.

The **exit( )** function in line 83 is used to immediately terminate the program. The status code, `EXIT_FAILURE`, is passed to the operating system to

对于任意供应商代码，将其减1并乘以sizeof(struct supplier_rec)，即可得到对应于该代码的文件记录的偏移量。

第83行的函数exit( )用于立即结束程序的执行，返回状态码EXIT_FAILURE给操作系统，表示程序执行失败。注意，这里不能使用return语句，因为return只能返回给主

indicate failure. Note that `return` cannot be used here because a `return` would only return to `main()`, and not terminate the program as required.

函数main( )，不能按预期的那样结束程序的执行。

13.12 Finding the position in a file using ftell() [使用函数ftell( )查找文件的当前位置]

The **ftell( )** function returns the offset value of the next byte that will be read or written in a file. The general format of `ftell()` is:

函数ftell( )主要用于返回文件中下一个将要读或写的字节的偏移量。

```
file_position = ftell( file_pointer ) ;
```

where `file_position` is an integer and `file_pointer` is a file pointer to an open file.

file_position是一个整型变量，file_pointer是一个指向已打开文件的文件指针。

Typically, you can use `ftell()` to store the current position in a file so that you can return to it at a later stage in the program.

函数ftell( )的典型应用是，保存文件的当前位置值，以便在程序的后续阶段再返回到这个位置。

You can also use `ftell()` along with `fseek()` to find the length of a file, as follows:

我们也可以综合使用函数ftell( )和fseek( )来计算文件的长度。

```
fseek( file_pointer, 0, SET_END ) ;   /* Go to end of file.*/
file_length = ftell( file_pointer ) ; /* The end position is
                                         the file length. */
```

13.13 Deleting a file using remove() [使用函数remove( )删除文件]

The next program demonstrates the deletion of a file using **remove()**. If the file is successfully deleted, `remove()` returns `0`; otherwise it returns `-1`.

Program Example P13L

```
1  /* Program Example P13L
2     To demonstrate file deletion using remove(). */
3  #include <stdio.h>
4  int main()
5  {
6    char file[81] ;
7    int return_code ;
8
9    printf( "Enter the name of the file to delete: ") ;
10   gets( file ) ;
11
12   return_code = remove( file ) ;
13
14   /* Has the file been successfully deleted? */
15   if ( return_code == 0 )
16     printf( "%s is deleted\n", file ) ;
17   else
18     printf( "%s cannot be deleted\n", file ) ;
19   return 0 ;
20 }
```

Programming pitfalls

1. When opening a file, always test the value of the file pointer for NULL. It will stop you trying to read from a file that does not exist!

1. 打开文件时，一定要检查文件指针的值是否为NULL，这样可以使我们避免从一个并不存在的文件中读取数据。

2. If you use mode "w" or "wb" for an existing file, the file contents are destroyed.

2. 如果使用“w”或“wb”方式打开一个已存在的文件，那么文件的内容将会被破坏。

3. To open a file and test for an error on opening you can use

```
if ( (fp = fopen(file, mode)) == NULL ) ...
```

not

```
if ( fp = fopen(file, mode) == NULL ) ...
```

You need the extra set of parentheses because the assignment operator = has a lower precedence than the comparison operator ==.

3. 可以使用下面的语句打开文件并检查文件打开过程中是否出现错误。
由于赋值运算符=的优先级低于关系运算符==，因此，在用上述语句进行相等比较前，需使用一个额外的圆括号将赋值表达式括起来。

4. File names frequently contain a \, for example, c:\newfile.dat. To open such a file you have to use an extra \ as in:

```
fp = fopen( "c:\\newfile.dat","r" ) ;
```

Without the extra \, the \n is interpreted as the newline escape sequence. Linux uses / in file specifications, and so the problem does not arise.

4. 文件的路径名中通常会包含反斜线\，例如c:\newfile.dat，在打开这样的文件时，必须多加一个反斜线\。
如果不多加一个反斜线\，那么\n将解释为换行符。在Linux操作系统中，使用/指定文件路径，因此不会出现此类问题。

5. Note that the while loop in program P13D has three sets of parentheses. All three sets are required. If you write the loop as

```
while ( char_in = fgetc(fp_in) != EOF )
  putchar( char_in ) ;
```

you will get a stream of characters with binary value 1 on the screen. This is because = is of lower precedence than !=. Therefore, the while loop given above is equivalent to:

```
while ( char_in = (fgetc(fp_in) != EOF) )
  putchar( char_in ) ;
```

In this while loop, the value of char_in will be the result of comparing the character read by fgetc(fp_in) with EOF, which will be true (a binary value of 1) or false (a binary value of 0), which is then assigned to char_in.

5. 注意，程序P13D中的while循环语句中有3对圆括号，这3对圆括号都是必须的。如果将该循环语句写成如下形式。
那么屏幕上将输出一连串的二进制值为1的字符。这是因为赋值运算符=的优先级低于关系运算符!=，因此这个while循环语句等价于下面的语句。
在这个while循环语句中，变量char_in的值是函数fgetc(fp_in)读取的字符与EOF进行比较的结果，该结果值要么为真（二进制值1），要么为假（二进制值0），这个结果值将赋值给变量char_in。

6. If a binary file is opened as a text file, it is quite possible that an end-of-file character EOF will be read from the file without the end of the file actually being reached.

6. 如果以文本文件方式打开一个二进制文件，那么很有可能将文件结束符EOF看做是普通数据从文件中读出，不会当成是读到了文件末尾。

7. In the `fread()` function, the first argument is a pointer to a memory location. The memory location must be big enough to store the block of memory being read.
8. If a file is opened for reading and writing, only reads or writes can be performed at any one time. To switch between reading and writing, or vice versa, you must first use `fseek()` or `rewind()`.

7. 函数fread( )的第一个实参是指向一个内存区域的指针，这个内存区域一定要足够大，以便能容纳得下函数要读取的数据块。
8. 如果文件是以只读或只写方式打开的，那么就只能对文件执行读或写操作，如果想在读写方式之间进行转换，那么必须先使用函数fseek( )或rewind( )重新定位文件的位置指针。

Quick syntax reference

	Syntax	Examples
Define a pointer to a file	`#include <stdio.h>` `FILE *file_pointer ;`	`FILE *fp ;`

Open a file	`file_pointer=`
	`   fopen(filename, mode) ;`
		`fp=fopen("file.txt","r")`
		`if(fp==NULL)`
		`{`
		`  printf("Open Error") ;`
		`  return EXIT_FAILURE ;`
		`}`

Read a character from a file	`variable=fgetc(file_pointer);`	`int char_in ;` `char_in=fgetc(fp) ;`
Write a character to a file	`fputc(variable, file_pointer);`	`int char_out ;` `fputc(char_out, fp) ;`

Read a character string from a file	`fgets(variable,`
	`      number_of_characters,`
	`      file_pointer) ;`
		`char str[10] ;`
		`fgets(str, 10, fp) ;`

Write a character string to a file	`fputs(variable, file_pointer);`	`fputs(str, fp) ;`
Text file formatted input	`fscanf(file_pointer,` `       format, variables) ;`	`int i ; float f ;` `fscanf(fp,"%d%f",&i,&f) ;`
Text file formatted output	`fprintf(file_pointer,` `    format,variables) ;`	`fprintf(fp, "%d %f", i, f) ;`

Binary file block input	`variable=`
	`fread(memory_pointer,`
	`      size, count,`
	`      file_pointer) ;`
		`fp=fopen("file.dat","rb");`
		`number_read=`
		`fread(&supp,sizeof(struct`
		`      supplier_rec),1,fp);`
		`if (number_read != 1)`
		`  printf("Error") ;`

(continued)

	Syntax	Examples
Binary file block output	`variable= fwrite(memory_pointer, size,count, file_pointer) ;`	`fp=fopen("file.dat","wb"); number_written= fwrite(&supp,sizeof(struct supplier_rec),1,fp); if (number_written!=1) printf("Error") ;`
Position a file pointer	`variable=fseek(file_pointer, offset,origin);`	`int offset ; int success ; ... success=fseek(fp,offset, SEEK_SET); if (success!=0) printf("Error") ;`
Position a file pointer to the start of a file	`rewind(file_pointer) ;`	`rewind(fp) ;`
Find the position of a file pointer	`offset=ftell(file_pointer) ;`	`offset=ftell(fp) ;`
Closing a file	`fclose(file_pointer) ;`	`fclose(fp) ;`
Deleting a file	`remove(file_name) ;`	`int success ; success=remove("me"); if (success == 0) printf("File deleted") ;`

Exercises

1. Write C statements to open the following files:

File name	File pointer	Mode	File type
supp.dat	supp_ptr	read	binary
cust.dat	cust_ptr	read and write	binary
temp.dat	fp	write	text
price.dat	price_file	append	text

2. The modes "a+", "r+" and "w+" allow for both reading and writing in a file. Which mode would you use for the following?

   (a) Updating existing data in a file.

   (b) Appending new data to the end of a file.

   (c) Deleting the contents of an existing file before writing new data to it.

3. What does this program do?

```
#include <stdio.h>
int main()
{
  int char_in ;
  FILE *fp_in ;

  if ( (fp_in = fopen("file.txt", "r")) != NULL )
```

```
  {
    while( (char_in = fgetc(fp_in)) != EOF )
    {
      if ( char_in > '/' && char_in < ':' )
        putchar( char_in ) ;
    }
  }
  fclose( fp_in ) ;
  return 0 ;
}
```

4. What is the difference between the following?

```
printf( "Message" ) ;
fprintf( stdout, "Message" ) ;
fprintf( stderr, "Message" ) ;
```

For Linux users:

What happens when you redirect the output with the Linux command

```
ex4 > temp
```

where `ex4` is a program containing the above three statements?

5. What is wrong with this program?

```
#include <stdio.h>
int main()
{
  int i ;
  FILE fp ;
  char *s = "test" ;
  if ( fp = fopen( "file", "r" ) != NULL )
  {
    for ( i = 0 ; i< 10 ; i++ )
    {
      fputs( fp, s ) ;
    }
  }
  fclose( "file" ) ;
  return 0 ;
}
```

6. Write your own version, `my_fgets()`, of the function `fgets()` using `fgetc()`.
7. Write your own version, `my_fputs()`, of the function `fputs()` using `fputc()`.
8. Write a program to compare two text files and display any differences between the files.
9. Write a program to remove blank lines from a text file.
10. Write a program to display the lines of a text file along with line numbers. The first line should be preceded by 1., the second by 2., and so on for each line in the file.

11. Write a program to create a text file containing payroll records with the following fields:

Field	Type	Example
Employee number (10000 to 50000)	unsigned int	12345
Employee surname	25 characters	Jones
Employee first name	10 characters	John
Hours worked (0 to 80)	int	4150 = 41.50
Pay rate (in pence)	int	1000
Percentage tax rate	int	2550 = 25.50%
Tax allowance(in pence)	int	4000

The data should be entered from the keyboard and written to the file. (Note that the data file will not contain decimals.)

12. Use the file created in exercise 11 to produce a report of the form:

```
Employee             <----- Hours -----> <------- Pay ------->
Number     Name      Normal   Overtime   Gross     Tax      Net
```

Overtime is time above 40 hours and is paid at time and a half. The program should total the gross, tax and net columns.

13. Write a program to display the contents of a text file on the screen, twenty lines at a time. The user will enter the file name. After every twenty lines are displayed the user will press Enter for the next twenty lines of the file. At the end of the file the number of lines in the file is displayed.

14. The Caesar cypher is a simple method for encrypting text. The cypher works by replacing each letter in the text with the letter that occurs a certain distance from it in the alphabet. Non-alphabetic characters are not encrypted. For example, ABcD is encrypted to BCdE when the distance is 1; xYZ1 is encrypted to zAB1 when the distance is 2.

    Write a program to encrypt a text file with a distance specified by the user.

15. Write a program to merge two text files. The two files contain names sorted in alphabetical order. The resultant merged file should also be sorted in alphabetical order.

Chapter Fourteen
The C Preprocessor
第 14 章　C 编译预处理

The preprocessor, as its name implies, processes a C program before it is read by the compiler. A C program is read by the preprocessor and modified according to preprocessor directives.

C program → preprocessor → modified C program → C compiler

Preprocessor directives are placed in a C program and tell the preprocessor how to modify a C program.

A preprocessor directive starts with a hash (#) in column one and is followed by a preprocessor directive. Some compilers allow you to have whitespace characters before and after the # to improve readability, but using this option may reduce the program's portability.

顾名思义，编译预处理就是指在编译器读取C程序之前先对C程序进行预处理。在程序编译之前，先由编译预处理程序读取C程序，然后根据编译预处理指令修改C程序。

C程序→编译预处理程序→修改的C程序→ C编译器。

C程序中的编译预处理指令，告诉编译预处理程序如何修改C程序。

编译预处理指令以井字号（#）开头，其后是编译预处理指令。某些编译器为了增加程序的可读性，允许在#前后添加空格，但这样会降低程序的可移植性。

14.1　Including files（包含文件）

The **#include** directive includes a specified file in the source file at the point at which the directive appears. The general format of the `#include` is:

#include编译预处理指令是将一个指定的文件包含到源文件中该指令所出现的位置。

```
#include <file_name>
```

or

```
#include "file_name"
```

The angle brackets < and > in the first format instruct the preprocessor to search for `file_name` in the standard include directory only. The standard include directory is where the standard include files such as `stdio.h` are stored.

The double quotation marks in the second format instruct the preprocessor to search for `file_name` in the *current* (where the program is) directory and, if it is not found there, to search in the standard include directory. However, if a directory is specified within the double quotation marks then this directory only is searched.

By convention, include files use the extension `.h` in their name.

Here are some examples of `#include` directives:

在第一种形式中，尖括号指示编译预处理程序仅在标准include目录中寻找包含的文件file_name，标准include目录是存储标准include文件（如stdio.h）的位置。

在第二种形式中，双引号指示编译预处理程序先在当前目录（C程序所在的目录）中寻找包含的文件，如果在当前目录中没有找到，那么再到标准include目录中寻找。但是，如果在双引号中指定了文件所在的目录，那么编译预处理程序将只在该目录中寻找包含的文件。

按照惯例，包含的文件一般以.h作为其文件的扩展名。

```
#include <stdio.h> /* Include stdio.h from standard include
                      directory.                               */
                      and if not there look for the file in
                      the standard include directory.          */
#include "\myinclude\my.h" /* Windows uses a single backslash \ */
#include "/myinclude/my.h" /* Linux uses a forward slash /      */
                    /* Include my.h from the directory
                       myinclude. Only the specified
                       directory myinclude is searched.        */
```

14.2 Defining macros（定义宏）

We have already used **#define** in previous programs to define symbolic constants, such as the number of elements in an array (see program P7C). The simplest format of `#define` is:

在前面的程序中，我们已经使用了#define来定义符号常量。

```
#define NAME replacement
```

where `NAME` is the symbolic constant. By convention, `NAME` is usually in uppercase characters. The preprocessor replaces all occurrences of `NAME` within the source file by `replacement` before the program is compiled. `NAME` is known as a *macro*, and the process of substituting a macro with its replacement text is called the *macro expansion*. `NAME` must conform to the rules for constructing valid C variable names.

其中，NAME为符号常量。按照惯例，符号常量通常采用大写字母来表示。在程序编译之前，编译预处理程序将源程序中所有出现NAME 的地方都替 换 为replacement。NAME称为**宏**，用替换文本取代宏的过程，称为**宏展开**，这里NAME必须遵循C变量的命名规则。

Examples:

```
#define MAX_ELEMENTS 10
#define DAYS_IN_WEEK 7
#define PI 3.141592653
/* You are not restricted to numeric constants. */
#define END_OF_SENTENCE '.'
#define DIGITS "0123456789"
#define FIVE_SPACES "  "
#define END_OF_STRING '\0'
#define NEWLINE '\n'
#define BACKSPACE '\b'
```

Once a macro has been defined, it cannot be assigned to a different value without first removing the original definition. This can be done with the **#undef** directive. For example:

宏一经定义，它就不能再指定为其他值，除非移除它的原始定义，我们可以使用#undef指令来实现这个功能。

```
#define N 100    /* N is 100.              */
#undef N         /* N is no longer defined. */
#define N 200    /* N is now 200.          */
```

You can use macros to replace elements of a C program, as shown in the next program.

在C程序中，可以使用宏来代替某些内容。

Program Example P14A

```
1  /* Program Example P14A
2     Demonstration of #define. */
```

```
3  #define BEGIN {
4  #define END }
5  #define IN_FORMAT "%d"
6  #define OUT_FORMAT "You entered the number %d\n"
7  #define PRINT printf
8  #define INPUT scanf
9  #define THEN )
10 #define IF if(
11 #include <stdio.h>
12 int main()
13 BEGIN
14   int n ;
15   PRINT( "Enter a number (0 to end) " ) ;
16   do
17   BEGIN
18     INPUT( IN_FORMAT, &n ) ;
19     IF n>0 THEN
20       PRINT( OUT_FORMAT, n ) ;
21   END
22   while ( n>0 ) ;
23   return 0 ;
24 END
```

What does this program do?

14.3 Macro parameters（带参数的宏）

The usefulness of a macro may be extended by the use of macro parameters. Using parameters in a macro allows you to vary the replacement text.

The next program defines a macro named SQUARE to calculate the square of a number.

在宏定义中使用参数可以扩展宏的实用性，通过在宏定义中使用参数，可以改变用于宏替换的文本。

下面的程序定义了一个名为SQUARE的宏，用于计算一个数的平方。

Program Example P14B

```
1  /* Program Example P14B
2     Demonstration of calling a macro with an argument. */
3  #define SQUARE(N)  (N)*(N)
4  #include <stdio.h>
5  int main()
6  {
7    int n = 2 ;
8    float f = 5.5 ;
9    int result ;
10
11   result = SQUARE( 2 ) ;
12   printf( "2 squared is %d\n", result ) ;
13   printf( "5.5 squared is %4.2f\n", SQUARE( f ) ) ;
14   result = SQUARE( n+1 ) ;
15   printf( "(2+1) squared is %d\n", result) ;
16   return 0 ;
17 }
```

The output from this program is:

```
2 squared is 4
5.5 squared is 30.25
(2+1) squared is 9
```

The macro `SQUARE` is defined with a parameter `N` in line 3. When the macro is called in line 11, the preprocessor replaces the right-hand side of the expression with the replacement text `(2)*(2)`. The compiler will therefore read line 11 as:

在程序第3行，宏SQUARE定义为带有一个形参N的宏，当宏在程序第11行第一次调用时，编译预处理程序将右侧的表达式SQUARE(2)替换为(2)*(2)。

```
result = (2)*(2) ;
```

The parameters of a macro are always enclosed in parentheses, to avoid side effects caused by the operator precedence rules. For example, line 14 currently expands to:

宏的参数必须用圆括号括起来，以避免因运算符优先级规则而产生的副作用。

```
result =(n+1)*(n+1) ;
```

Without the parentheses around each parameter in line 3, line 14 would expand to:

```
result = n+1 * n+1 ;
```

As `n` is `2`, `result` will now have a value of 5 (= 2 + 1*2 + 1) rather than 9. In fact there is still a bug lurking within this program. If line 14 is changed to

```
result = 90/SQUARE(n+1) ;
```

Will `result` have the value 10 after this statement is executed? No, because the preprocessor will expand this statement to

```
result = 90/(n+1)*(n+1) ;
```

giving `result` a value of 90 / 3 * 3 = 90, as `n` has `a` value of `2`. An extra pair of parentheses in line 3 will solve this problem; so the macro now becomes:

```
#define SQUARE(N) ((N)*(N)) ;
```

and

```
result = 90/SQUARE(n+1) ;
```

expands to

```
result= 90/((n+1)*(n+1)) ;
```

giving `result` the expected value of 90 / (3 * 3) = 10.

A general rule, therefore, is to use parentheses around each macro parameter and around the macro replacement as a whole. However, even these precautions cannot help if line 14 is:

因此，宏定义的一般原则是，为每个宏形参都添加一对圆括号，然后再为整个宏替换文本添加一对圆括号。但是，即使注意了这些事项，某些错误仍然是无法避免的。

```
result = SQUARE( n++ ) ;
```

Now line 14 will expand to

```
result = ((n++)*(n++)) ;
```

which, as `n` is 2, gives `result` a value of:

`result` = 2 * 3 = 6, and n gets incremented twice to 4.

The only solution here is to avoid pre-operators and post-operators as arguments in a macro.

避免在带参数的宏定义中使用前置或后置运算符。

14.4 Macros and functions（宏和函数）

Although they look very alike, macros differ from functions in a number of important ways:

1. The data type of macro parameters are not specified, so the arguments used in calling the macro can be of any data type. For example, line 11 of program P14B called the macro `SQUARE` with an integer argument, and line 13 called the same macro with a floating-point argument. In contrast, a parameter in a function can be of one data type only.

   宏可以带任何类型的参数。例如，程序P14B第11行中宏SQUARE的参数为整型，而第13行中宏SQUARE的参数又变成了浮点型。与此相反，函数参数的类型只能是一种数据类型。

2. A call to a function has processing overheads associated with it, for example passing a copy of the arguments, saving the return address, allocating storage for local variables, and so on. These overheads are not present when a macro is used, because the macro has been expanded at the preprocessing stage.

   函数调用需要额外的相关处理开销，如传递实参的副本，保存函数的返回地址，为局部变量分配内存空间等。但是，使用宏就不需要这些处理开销，因为宏是在程序的编译预处理阶段被扩展的。

3. Every time a macro is expanded the code it produces is placed in the program; so using a macro ten times will result in ten expansions. With a function, the code only appears once in a program, regardless of how many times the function is used.

   在程序中的每个应用宏的地方都需要对宏进行扩展，因此应用10次宏就需要10次宏扩展。而对于函数，无论函数在程序中被调用多少次，其代码只在程序中出现1次。

4. A function has not got the same side effects as a macro, as previously discussed in conjunction with the macro `SQUARE`. For example, instead of the macro `SQUARE` you could use the following equivalent function:

   ```
   int square( int num )
   {
     return ( num * num ) ;
   }
   ```

   A statement such as `result = square(n++)` will correctly place `n*n` in `result` and increment `n`. Using the macro `SQUARE` you get `(n)*(n+1)` assigned to `result`, and `n` is incremented twice.

   如前所述，使用宏SQUARE的时候会产生一些副作用，但使用函数的时候，就不会出现这些副作用。例如，若使用与上述宏SQUARE等价的函数来替代它。那么像result = square(n++)这样的语句使变量result可以得到正确的计算结果n*n，同时变量n也只会自增一次。而使用宏SQUARE时，变量result的值却赋值为(n)*(n+1)，同时变量n自增了两次。

5. Some compilers limit a macro definition to one line; there is no limit to the number of lines in a function.

   某些编译器限定，宏定义只能写在一行上。而编译器对函数的行数却是没有限定的。

14.5 Some useful macros（一些有用的宏）

Below are some macros that you will find useful in your C programs. You could type any of these macros into a C program as the need arises. Alternatively, type them into a file (e.g. macros.h) and `#include` this file at the start of a program.

下面列出的一些宏定义在C程序中是很有用的，可以在需要的时候在C程序中使用这些宏，或者将它们放入一个文件（例如macros.h）中，然后在程序的开始处用#include将该文件包含进来。

Macro:
```
/* Find minimum of two numbers. */
#define MIN(A, B) ((A)<(B)?(A):(B))
```
Example:
```
min_val = MIN( n1, n2 ) ;
```

Macro:
```
/* Find maximum of two numbers. */
#define MAX(A, B) ((A)>(B) ? (A):(B))
```
Example:
```
max_val = MAX( n1, n2 ) ;
```

Macro:
```
/* Find absolute value of a number. */
#define ABS(N) ((N)<0 ? -(N):(N))
```
Example:
```
abs_val = ABS(n) ;
```

Macro:
```
/* Check whether a character is a + or - */
#define ISSIGN(C) ((C)=='+'||(C)=='-' ? 1:0)
```
Example:
```
if ( ISSIGN(char_in) )
  printf( "%c is a sign", char_in ) ;
```

Macro:
```
/* Check whether a variable falls between two values. */
#define INRANGE(V, L, U) ((V)>=(L)&&(V)<=(U) ? 1:0)
```
Example:
```
if ( INRANGE(n, lower_val, upper_val) )
  printf( "n is between %d and %d", lower_val, upper_val ) ;
```

14.6 Conditional directives（条件编译预处理指令）

The preprocessor conditional directives **#ifdef** and **#endif** are used to include or omit blocks of C statements. For example, suppose when debugging a program you want to display the values of three integer variables `a`, `b` and `c` at a certain point in the program. Place the following in your program at the point where you want to display the variables:

条件编译预处理指令#ifdef和#endif用于包含或忽略某段C语句。

```
#ifdef DEBUG
printf( "a, b and c at this point are: %d %d %d", a, b, c ) ;
#endif
```

If `DEBUG` has previously been defined with

```
#define DEBUG
```

the `printf()` statement will be included in your program. If you leave out this `#define` or have `DEBUG` undefined with

```
#undef DEBUG
```

the `printf()` statement will not be placed in the program. Debugging statements can be included or excluded in your program, therefore, simply by defining or undefining `DEBUG`.

使用条件编译预处理指令，通过定义DEBUG或取消DEBUG的定义，可以将一些调试用的语句包含到程序中，或者从程序中移除调试用的语句。

14.7 Character-testing macros（字符检测宏）

The standard header file `ctype.h` contains a set of macros that can be used to test the value of a single `character`. These macros, as shown in Table 14.1, return a true (non-zero integer) value or a false (zero integer) value depending on whether or not the character belongs to a particular set of characters.

在标准头文件ctype.h中包含了用于检测单字符值的一个宏集合。这些宏如表14.1所示，它们首先判断字符是否属于某个特定的字符集，然后据此返回真值（即非零整数值）或假值（即零值）。

Table 14.1 character-testing macros

Macro	Character set
isalnum	Alphanumeric character: A–Z, a–z, 0–9.
isalpha	Alphabetic character: A–Z, a–z.
isascii	ASCII character: ASCII codes 0–127.
iscntrl	Control character: ASCII codes 0–31 or 127.
isdigit	Decimal digit: 0–9.
isgraph	Any printable character other than a space.
islower	Lowercase letter: a–z.
isprint	Any printable character, including a space.
ispunct	Any punctuation character.
isspace	Whitespace character: \t, \v, \f, \r, \n or space ASCII codes 9–13 or 32.
isupper	Uppercase letter: A–Z.
isxdigit	Hexadecimal digit: 0–9 and A–F.

Each of the macros in Table 14.1 uses an integer or character parameter and returns an integer value of zero (logical false) if the argument value is not in the set or non-zero (logical true) if the argument is contained in the set.

Note: You should use `isascii()` to verify that the argument is a valid ASCII character before using any of these macros.

表14.1中的每个宏都带有一个整型或字符型的参数。若宏参数属于指定的字符集，则返回一个非零值（逻辑真），否则返回零值（逻辑假）。

注意：在使用上述宏之前，应使用isascii( )检验一下参数字符是否为有效的ASCII码字符。

Example:

```
#include <ctype.h>    /* ctype.h must be included. */
#include <stdio.h>
nt main()
{
  char ch ;
  ...
```

```
  if ( isascii(ch) )
  {
    if ( isupper(ch) )
      printf( "%c is an uppercase character\n", ch ) ;
    ...
  }
  return 0 ;
}
```

The header file `ctype.h` contains two further macros: `tolower` and `toupper`. (See Table 14.2.)

Table 14.2 `tolower` and `toupper` macros

Macro	Purpose
`tolower`	Converts an uppercase character to lowercase.
`toupper`	Converts a lowercase character to uppercase.

Example:

```
#include <ctype.h>
#include <stdio.h>
int main()
{
  char ch = 'a' ;

  printf( "Convert to uppercase %c", toupper(ch) ) ;
  printf( "Back to lowercase %c", tolower(ch) ) ;

  return 0 ;
}
```

14.8 The `assert()` macro [`assert()` 宏]

The `assert()` macro is used to identify logic errors ("bugs") in a program during development.

This macro uses a single argument which can be a variable or an expression. If the argument evaluates to logical true (a binary value of 1), `assert()` does nothing; otherwise an error message is displayed on `stderr` (standard error stream to display error messages – see Table 13.2) and the program stops.

The next program reads in values from a text file into an array and then displays the values in the array. Each function in the program uses `assert()` to test the assumptions made by the function, e.g. the file exists.

assert()宏用于在程序开发调试过程中识别程序中的逻辑错误。

使用assert()宏只需要一个实参，这个实参可以是变量或者表达式。如果这个实参的值是逻辑真（例如二进制值1），那么assert()什么也不做，否则在标准错误输出设备（标准错误流显示错误信息参见表13.2）上显示一个错误信息，并且终止程序的执行。

```
1   /* Program example P14C
2      Demonstration of the assert macro. */
3    #include <assert.h> /* Required header file */
4    #include <string.h>
5    #include <stdio.h>
6    # define MAX_SIZE 9
7    # define FILE_NAME "numbers.txt"
```

```
8     void display_array( int a[], int n );
9     int file_to_array(char* file_name,int array[],int max_array_size);
10
11    int main()
12    {
13      int numbers[ MAX_SIZE ] ;
14      int number_of_elements ;
15
16     /* Read integers from a text file into an array. */
17     number_of_elements = file_to_array(FILE_NAME, numbers, MAX_SIZE);
18     /* Display the array. */
19     display_array( numbers, number_of_elements );
20
21     return 0;
22    }
23
24    /* Function   : file_to_array
25       Purpose    : Read integers from a text file into an array.
26       Parameters: The file name, the array and the size of the array.
27       Returns    : The number of integers stored in the array.       */
28    int file_to_array(char* file_name,int array[],int max_array_size)
29    {
30      int num_elements=0;
31      int input_value;
32      FILE *fp;
33
34      fp = fopen( file_name, "r" ) ;
35      /* Assert that the file has opened successfully. */
36      assert( fp != NULL ) ;
37      /* Read in the values from the file into the array. */
38      while( (fscanf(fp,"%d",&input_value)) != EOF )
39      {
40        /* Assert that the array is not too small. */
41        assert ( num_elements < MAX_SIZE ) ;
42        array[ num_elements ] = input_value;
43        num_elements++ ;
44      }
45      fclose(fp) ;
46
47      return num_elements ;
48    }
49
50    /* Function   : display_array
51       Purpose    : Display the values in an integer array.
52       Parameters: The array and the number of elements in the array.
53    */
54    void display_array( int array[], int n )
55    {
56      int i ;
57    /* Assert the number of elements in the array is > 0 *1
```

```
58     assert ( n > 0 ) ;
59     for ( i=0; i<n; i++ )
60       printf( "%d ", array[i] ) ;
61     printf( "\n" );
62   }
```

When this program is executed the following nine values are displayed:

```
78 12 87 11 32 40 29 6 1951
```

To see how `assert()` catches errors do either of the following:

- Reduce the maximum size of the array on line 6 to less than 9. This will make the array too small to store all the vlues in the file.
- Change the name of the file in line 7 to an non-existent file name.
- Change the name of the file to an existing file that has no values in it.

For example, reducing the maximum size of the array on line 6 results in a message similar to the following:

```
Assertion failed: (num_elements < MAX_SIZE), function
file_to_array, file p14c.c, line 41.
```

The message indicates where the bug is in the program. It is important to note that this message is intended for the programmer and not for an end-user of the program.

Once, program is thoroughly tested, assertion checking can be turned off by defining the `NDEBUG` before line 3:

```
#define NDEBUG
#include <assert.h>
```

To turn assertion checking back on, remove `#define NDEBUG`.

为了观察assert()是如何捕获错误的，可以做下面任何一项工作：

- 将程序第6行的数组大小减小到小于9，这将使得数组太小无法保存文件中的数据；
- 修改程序第7行的文件名为一个不存在的文件名；
- 改文件名为一个已存在的文件名，但是这个文件中没有任何内容。

Programming pitfalls

1. Avoid side effects caused by macros by placing the entire macro replacement text and each macro parameter in parentheses. (See the text for a discussion of side effects.)
2. Use `isascii()` before using any of the other `is` macros. These macros are valid only when the argument is an ASCII value or `EOF`.
3. To `#include` a header file such as `c:\sub_dir\file.h` you use the preprocessor directive

```
#include "c:\sub_dir\file.h"
```

Note that a double backslash is not required with `#include`. Contrast this with programming pitfall 4 in Chapter Thirteen, where a double backslash is required with `fopen()`.

1. 为了避免宏引起的副作用，我们需要将整个宏替换文本及每个宏形参都用圆括号括起来（参见前面对宏的副作用的讨论）。
2. 在使用其他is宏之前，应先使用isascii( )检测一下参数是否为有效的ASCII 码字符。仅在参数为有效的ASCII码字符或EOF时，这些宏才有效。
3. 如果要包含头文件c:\sub_dir\file.h，则应使用下面的编译预处理指令。
注意，此时并不需要使用双反斜线\\，与第13章编程注意事项第4条是不同的，在fopen( )函数中必须使用双反斜线。

Quick syntax reference

	Syntax	Examples
Define a symbolic constant	`#define IDENTIFIER value`	`#define PI 3.14`
Define a macro	`#define NAME(parameter(s)) (body)`	`#define SQ(N) ((N)*(N))`

Exercises

1. Write `#define` directives to do the following:
   (a) replace the equality symbol == with the word `EQUALS`
   (b) replace the space character `' '` with the word `SPACE`
   (c) replace the word `char` with the word `BYTE`
   (d) replace `&&` with the word `AND`
   (e) replace `||` with the word `OR`.
2. Write a macro `DISPLAY(A)` to display the name, value and address of a variable.
3. Write a macro `ADD(A, B)` to add two numeric values.
4. Write a macro `NEG(N)` that returns the negative value of its argument.
5. Write a macro `DIV(A, B)` that returns `A/B` when `B` is not 0 and otherwise returns 0.
6. Write a macro `IS_EVEN(N)` that returns `1` if `N` is an even number and otherwise returns 0.
7. Write a macro `IS_IN_ORDER(A, B, C)` that returns 1 if A, B and C are in increasing order; otherwise it returns 0.
8. Write a macro `TRUNCATE(N)` to truncate the fractional part of a number. For example, `TRUNCATE(3.15)` is 3 and `TRUNCATE(3.95)` is also 3.
9. Write a macro `TOINT(c)` to convert an ASCII digit to its integer equivalent.

10. Write a macro `REM(A, B)` which takes two integer arguments and returns the remainder when the first integer is divided by the second.
11. Write a macro `SWAP(A, B)` which swaps two integer arguments.
12. Modify the solution to exercise 11 to swap two arguments with any data type, not just integers. For example, `SWAP(float, f1, f2)` where `f1` and `f1` are `floats`.
13. Write a function to convert all uppercase characters in a string to lowercase.
14. Write a function to convert all non-numeric characters in a string to spaces.
15. Write a function to calculate the number of digits, the number of alphabetic characters and the number of control characters in a string.

Appendix A
List of C Keywords

auto	double	int	struct
break	else	long	switch
case	enum	register	typedef
char	extern	return	union
const	float	short	unsigned
continue	for	signed	void
default	goto	sizeof	volatile
do	if	static	while

In addition to this list, consider all names beginning with an underscore (_) to be reserved for system use.

C99 add five new keywords：`inline`、`restrict`、`_Bool`、`_Complex`、`_Imaginary`.

C11 add one new keyword：`_Generic`.

Appendix B
Precedence and Associativity of C Operators

Precedence	Operator	Name	Associativity
15	( )	Parentheses	Left to right
	[ ]	Subscripts	
	.	Dot	
	->	Arrow	
	++	Postfix increment	
	--	Postfix decrement	
14	-	Unary minus	Right to left
	+	Unary plus	
	~	Ones complement	
	!	Logical NOT	
	*	Indirection or dereference	
	&	Address	
	++	Prefix increment	
	--	Prefix decrement	
	sizeof	Size of	
	(type)	Type cast	
13	*	Multiply	Left to right
	/	Divide	
	%	Modulus	
12	+	Add	Left to right
	-	Subtract	
11	<<	Left shift	Left to right
	>>	Right shift	
10	<	Less than	Left to right
	>	Greater than	
	<=	Less than or equal to	
	>=	Greater than or equal to	
9	==	Equal to	Left to right
	!=	Not equal to	
8	&	Bitwise AND	Left to right
7	^	Bitwise exclusive OR	Left to right
6	|	Bitwise OR	Left to right
5	&&	Logical AND	Left to right
4	||	Logical OR	Left to right
3	? :	Conditional	Right to left

(condinued)

Precedence	Operator	Name	Associativity
2	=	Assignment	Right to left
	*=	Multiply and assign	
	/=	Divide and assign	
	%=	Modulus and assign	
	+=	Add and assign	
	-=	Subtract and assign	
	<<=	Left shift and assign	
	>>=	Right shift and assign	
	&=	Bitwise AND and assign	
	|=	Bitwise OR and assign	
	^=	Bitwise exclusive OR and assign	
1	,	Comma (sequence)	Left to right

Notes:

1. Expressions within parentheses have a higher precedence than expressions without parentheses.
2. When an expression contains several operators with equal precedence, evaluation proceeds according to the associativity of the operator, either from right to left or from left to right.

Appendix C
ASCII Character Codes

A character code is a numerical value used to represent a character in the computer's memory. The ASCII (American Standard Code for Information Interchange) character set is defined as a table of seven-bit codes that represent control characters and printable characters. For example, the letter "A" is represented by ASCII code 65, and "1" is represented by ASCII code 49.

The circumflex symbol ^ is used in the following table to indicate that the control key Ctrl is pressed simultaneously with another key.

Decimal code	Hexadecima lcode	Key	Decimal code	Hexadecima lcode	Key
0	00	^@	27	1B ESC	^[
1	01	^A	28	1C	^\
2	02	^B	29	1D	^]
3	03	^C	30	1E	^^
4	04	^D	31	1F	^_
5	05	^E	32	20	space
6	06	^F	33	21	!
7	07 BEL	^G	34	22	"
8	08 BS	^H	35	23	#
9	09 TAB	^I	36	24	$
10	0A LF	^J	37	25	%
11	0B VT	^K	38	26	&
12	0C FF	^L	39	27	'
13	0D CR	^M	40	28	(
14	0E	^N	41	29	)
15	0F	^O	42	2A	*
16	10	^P	43	2B	+
17	11	^Q	44	2C	,
18	12	^R	45	2D	–
19	13		46	2E	.
20	14	^T	47	2F	/
21	15		48	30	0
22	16		49	31	1
23	17	^W	50	32	2
24	18	^X	51	33	3
25	19	^Y	52	34	4
26	1A	^Z	53	35	5
54	36	6	91	5B	[
55	37	7	92	5C	\
56	38	8	93	5D	]
57	39	9	94	5E	^
58	3A	:	95	5F	_
59	3B	;	96	60	`
60	3C	<	97	61	a

(condinued)

<table>
<tr><th>Decimal code</th><th>Hexadecima lcode</th><th>Key</th><th>Decimal code</th><th>Hexadecima lcode</th><th>Key</th></tr>
<tr><td>61</td><td>3D</td><td>=</td><td>98</td><td>62</td><td>b</td></tr>
<tr><td>62</td><td>3E</td><td>></td><td>99</td><td>63</td><td>c</td></tr>
<tr><td>63</td><td>3F</td><td>?</td><td>100</td><td>64</td><td>d</td></tr>
<tr><td>64</td><td>40</td><td>@</td><td>101</td><td>65</td><td>e</td></tr>
<tr><td>65</td><td>41</td><td>A</td><td>102</td><td>66</td><td>f</td></tr>
<tr><td>66</td><td>42</td><td>B</td><td>103</td><td>67</td><td>g</td></tr>
<tr><td>67</td><td>43</td><td>C</td><td>104</td><td>68</td><td>h</td></tr>
<tr><td>68</td><td>44</td><td>D</td><td>105</td><td>69</td><td>i</td></tr>
<tr><td>69</td><td>45</td><td>E</td><td>106</td><td>6A</td><td>j</td></tr>
<tr><td>70</td><td>46</td><td>F</td><td>107</td><td>6B</td><td>k</td></tr>
<tr><td>71</td><td>47</td><td>G</td><td>108</td><td>6C</td><td>l</td></tr>
<tr><td>72</td><td>48</td><td>H</td><td>109</td><td>6D</td><td>m</td></tr>
<tr><td>73</td><td>49</td><td>I</td><td>110</td><td>6E</td><td>n</td></tr>
<tr><td>74</td><td>4A</td><td>J</td><td>111</td><td>6F</td><td>o</td></tr>
<tr><td>75</td><td>4B</td><td>K</td><td>112</td><td>70</td><td>p</td></tr>
<tr><td>76</td><td>4C</td><td>L</td><td>113</td><td>71</td><td>q</td></tr>
<tr><td>77</td><td>4D</td><td>M</td><td>114</td><td>72</td><td>r</td></tr>
<tr><td>78</td><td>4E</td><td>N</td><td>115</td><td>73</td><td>s</td></tr>
<tr><td>79</td><td>4F</td><td>O</td><td>116</td><td>74</td><td>t</td></tr>
<tr><td>80</td><td>50</td><td>P</td><td>117</td><td>75</td><td>u</td></tr>
<tr><td>81</td><td>51</td><td>Q</td><td>118</td><td>76</td><td>v</td></tr>
<tr><td>82</td><td>52</td><td>R</td><td>119</td><td>77</td><td>w</td></tr>
<tr><td>83</td><td>53</td><td>S</td><td>120</td><td>78</td><td>x</td></tr>
<tr><td>84</td><td>54</td><td>T</td><td>121</td><td>79</td><td>y</td></tr>
<tr><td>85</td><td>55</td><td>U</td><td>122</td><td>7A</td><td>z</td></tr>
<tr><td>86</td><td>56</td><td>V</td><td>123</td><td>7B</td><td>{</td></tr>
<tr><td>87</td><td>57</td><td>W</td><td>124</td><td>7C</td><td>|</td></tr>
<tr><td>88</td><td>58</td><td>X</td><td>125</td><td>7D</td><td>}</td></tr>
<tr><td>89</td><td>59</td><td>Y</td><td>126</td><td>7E</td><td>~</td></tr>
<tr><td>90</td><td>5A</td><td>Z</td><td>127</td><td>7F</td><td>Del</td></tr>
</table>

Appendix D
Fundamental C Built-in Data Types

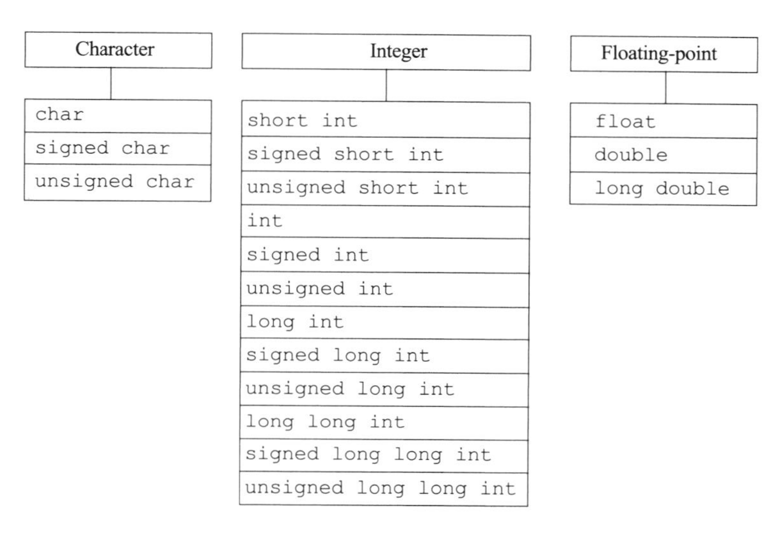

D.1 Integer constants

Normally an integer constant is specified as a decimal (base 10) number, but it can also be specified in octal (base 8) or hexadecimal (base 16).

An octal number must start with a 0 and a hexadecimal number must start with either `0x` or `0X`.

Decimal numbers must not start with a 0.

All the following statements assign decimal 65 to the integer variable `v1` as defined on line 6 of program P2C.

```
v1 = 65 ;   /* Decimal 65 is assigned to v1 */
v1 = 0101 ; /* Octal 101 = decimal 65 is assigned to v1 */
v1 = 0x41;  /* Hexadecimal 41 = decimal 65 is assigned to v1 */
```

An integer value can be displayed in octal or hexadecimal using `%o` or `%x` as the format specifier in `printf()`.

D.2 Floating-point constants

A floating point constant is normally specified in the decimal base.

A floating-point constant must have a decimal point or an `e` (or `E`) with an exponent or both. The exponent is a power of 10.

All the following statements assign decimal `-18.23` to the floating-point variable `v2` as defined on line 7 of program P2C.

```
v2 = -18.23 ;
v2 = -1823e-2 ;
v2 = -.1823e+2 ;
```

A floating-point value can be displayed in exponential form by using the `%g` or `%G` format specifiers in `printf()`.

D.3 Input-Output Format Specifiers for the Built-in Data Types

Table D.1 commonly used `printf()` and `scanf()` format specifiers

Data type	`printf ()`	`scanf ()`
`char`	`%c`	`%c`
`short int` `signed short int`	`%d`	`%d`
`unsigned short int`	`%u`	`%u`
`int` `signed int`	`%d`	`%d`
`unsigned int`	`%u`	`%u`
`long int` `signed long int`	`%ld`	`%ld`
`unsigned long int`	`%lu`	`%lu`
`float`	`%f`	`%f`
`long long int` `signed long long int`	`%I64d` or `%lld`	`%I64d` or `%lld`
`unsigned long long int`	`%I64u` or `%llu`	`%I64u` or `%llu`
`double`	`%f` or `%lf`	`%lf`
`long double`	`%Lf`	`%Lf`

Note:

`%ld` `%lu` `%lld` `%llu` — `l` is lowercase L

`%I64d` `%I64u` — `I` is uppercase i